Mathcad for
Chemical Engineers

Second Edition

Hertanto Adidharma
University of Wyoming

Valery Temyanko
University of Arizona

Order this book online at www.trafford.com
or email orders@trafford.com

Most Trafford titles are also available at major online book retailers.

Note for Librarians: A cataloguing record for this book is available from Library
and Archives Canada at www.collectionscanada.ca/amicus/index-e.html

Printed in Victoria, BC, Canada.

ISBN: 978-1-4269-0812-5 (soft)
ISBN: 978-1-4269-0814-9 (ebook)

*We at Trafford believe that it is the responsibility of us all, as both individuals and corporations,
to make choices that are environmentally and socially sound. You, in turn, are supporting this
responsible conduct each time you purchase a Trafford book, or make use of our publishing services.
To find out how you are helping, please visit www.trafford.com/responsiblepublishing.html*

*Our mission is to efficiently provide the world's finest, most comprehensive book publishing
service, enabling every author to experience success. To find out how to publish your book, your
way, and have it available worldwide, visit us online at www.trafford.com*

Trafford rev. 5/27/2009

 www.trafford.com

North America & international
toll-free: 1 888 232 4444 (USA & Canada)
phone: 250 383 6864 ♦ fax: 250 383 6804 ♦ email: info@trafford.com

About the Authors

Hertanto Adidharma is currently a professor of chemical engineering at the University of Wyoming. He has more than ten years experience in chemical engineering education. He has taught various chemical engineering courses: Material and Energy Balance, Transport Phenomena, Chemical Reaction Engineering, Thermodynamics, Numerical Methods, Computer Programming, Unit Operations, and Introduction to Chemical Engineering Computing. Before entering teaching, he spent several years in chemical industry and became an independent consultant.

Valery Temyanko is currently a Staff Engineer, Sr. at the University of Arizona. He has more than 10 years of industrial experience. Research and development of resin transfer molding (RTM), reaction injection molding (RIM), reforming of fuel, process optimization and control are the primary areas of his expertise.

Dedicated to:

The loves of my life:

My mother, *Poo Lee Djoe*, and the memory of my father, *Tan Boen Hwa*,
My wife, *Mariwati Hardjito*,
My daughters, *Widya* and *Lingga*.

Hertanto Adidharma

My beloved family:

My mother, *Vilora Temyanko*, and my father *Leonid Temyanko*,
My wife, *Elena Temyanko*,
My daughters, *Svetlana*, *Lora*, and *Sophia*.

Valery Temyanko

Contents

Preface

This book serves as a must-have guide and quick reference for chemical engineers and those who would like to learn and use Mathcad as their computational tool. The book is also intended for one-semester course on Chemical Engineering Computing using Mathcad. Authors believe that unlike other computational software packages, Mathcad is very intuitive and thus easy to learn.

There are several mathematical or computational software packages available today. Among the most widely used are Maple, Maxima, MatLab, Mathematica, Mathcad, and Excel. The preference of one particular computational package over the others is usually, ironically, based only on familiarity. The approach of "familiar" software works with some limitations, but it is never the most efficient way of problem solving. It is very important to choose and use the latest software and not to be afraid of switching and learning. This is the best time investment that one can do. Given many years of experience in solving a wide variety of chemical engineering tasks, authors are convinced that Mathcad is easy to learn and easy to use, yet versatile and powerful. Thus, it is an extremely important and useful software package for modern chemical engineers. Although Mathcad 13 is the software package chosen by the authors and will be used throughout the book, most of the features discussed can also be applied using earlier versions of Mathcad. Also, although Mathcad will always evolve into a newer version, most of the contents in this book will be applicable for any subsequent version of Mathcad.

The main objective of the book is to develop computational skills needed in chemical engineering applications. The readers will learn how to solve different types of mathematical models. At the same time, the book introduces readers to many chemical engineering applications, such as

- Property estimation
- Material and energy balance
- Thermodynamics
- Transport phenomena
- Kinetic and reactor design
- Unit operations

- Process control
- Process economics
- Statistical thermodynamics

The already-developed mathematical models in those applications will be solved using Mathcad. The development of the mathematical models itself, from fundamental theories, is beyond the scope of this book. Readers must consult other chemical engineering textbooks. For students, they usually learn how to develop such models from other courses.

In this book, readers can learn how to use Mathcad functions in clearly described procedures intended for quick reference. To introduce readers to some related mathematical problems appeared in chemical engineering applications, examples and problems are given in each chapter.

Finally, it is important to mention that the book is not intended to give readers a complete manual of Mathcad, but rather to provide them with the essential Mathcad functions and features commonly needed in chemical engineering. After reading this book, of course the readers are encouraged to keep exploring the power of Mathcad.

Hertanto Adidharma

Valery Temyanko

Preface to the Second Edition

Although most of the materials in the first edition of our book are quite general and applicable to the latest version of Mathcad, we believe that the book needs to be updated to reflect the new features of Mathcad and to improve its presentation.

In this second edition, Mathcad 14 is used. We implement inline numerical calculation in the procedures and examples, which improves worksheet organization and clarity. Chapter 2 is revised to enable new Mathcad users (beginners) to learn completely on their own. The description of *find* function in Chapter 4 is revised to reflect the feature of the function in Mathcad 14. The description of interpolation in Chapter 5 is revised and made clearer. The description of *odesolve* function in Chapter 8 has also been revised to reflect the change in the default solver implemented in Mathcad 14. Furthermore, we add more problems to give readers more exposure to chemical engineering applications.

We thank our students for the numerous questions for clarification leading to several improvements and typo corrections. We also benefit substantially from the advice and encouragement of our colleagues in our respective departments, who are too numerous to acknowledge individually.

We believe that the book will continue to serve as a good reference for chemical engineering students and chemical engineers, who would like to learn and use Mathcad as their computational tool, and a good textbook for a course on chemical engineering computing.

Hertanto Adidharma

Valery Temyanko

Mathcad for
Chemical Engineers

Second Edition

Chapter 1
Introduction

The profession of chemical engineering officially started as early as 1888, when Professor Lewis Norton of the Massachusetts Institute of Technology introduced "Course X" and united chemical engineers with a formal degree. Soon after that, Tulane University and the University of Pennsylvania followed the suit by adding four-year program to their curriculum. The first programs were the combination of mechanical engineering and chemistry with an emphasis on the industrial needs of that time. Indeed, the challenges at that time required very different logics compared to what modern chemical engineers are facing today. The complexity of the logics grew as the process requirement became more demanding due to economical, environmental, and/or technological considerations.

Nowadays, in performing their tasks, such as operation, equipment design, process design, process control, and process analysis, chemical engineers utilize more sophisticated mathematical frameworks. As shown in Figure 1.1, the problem solving in chemical engineering involves several steps. First, the physical problem with all of the constraints must be clearly defined. At this stage, it is also very important to properly state the final goal. Second, a mathematical model of the system of interest has to be formulated using fundamental theories such as force balance, momentum, heat, and mass balance, thermodynamics, and rate laws (momentum, heat, and mass transfer, kinetics). This is a critical step in problem solving. If a model is incorrectly formulated, one will obtain erroneous results although the mathematical solution of the model is perfect.

Third, after a mathematical model has been correctly formulated, the task is reduced to how to solve it fast and reliably. Before the computer era, mathematical models had to be simple enough so that analytical or numerical solutions can be obtained in a reasonable period of time. Since numerical solutions in the past were very time consuming, if achievable at all, chemical engineers sacrificed the accuracy of the solutions by oversimplifying the problems. Fortunately, modern computing allows us to perform such tasks much faster and deal with more and more complex mathematical formulations. This means that unnecessary assumptions in developing mathematical models could be removed, optimization of an integrated system could be performed, and what-if analysis could be easily done. When a numerical method or procedure is developed, it must be carefully and rigorously checked for errors and tested on some problems of known solutions.

Due to the advanced improvement on the mathematical software packages, many basic numerical methods, such as routines (functions) for solving non-linear equations and ordinary differential equations, have been at our fingertips. Thus, the errors originated from numerical methods used can be minimized.

Fourth, after a mathematical solution has been obtained, result verification must be performed to ensure the correctness of the problem solution. This can be done by critically analyzing the results and/or comparing with the available experimental data. Of course, in analyzing the results, logical thinking, good sense, and deep understanding of the fundamental theories underlying the physical problem are needed. If the results violate theories, do not make sense, or do not agree with experimental data, a revision of the model or numerical approach should be considered.

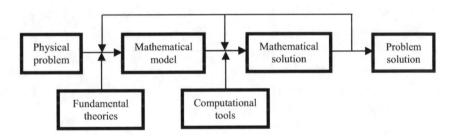

Figure 1.1. Problem solving in chemical engineering

The success of problem solving in chemical engineering is clearly determined by *the accuracy of the model, the versatility of the computation, and the capacity of analysis*. The first factor calls for a thorough understanding of the fundamental theories underlying the physical problem, the second calls for a versatile and powerful computational tool, and the third calls for both understanding theories and computational tool. Therefore, learning a computational tool is critical in chemical engineering.

We need a computational tool for a certain problem not only to solve the mathematical model obtained, but also to learn the behavior of the system, for which what-if analysis is usually performed. We intuitively like to experiment, by varying parameters and variables and analyzing their effects, to obtain a better understanding of the system behavior. Learning by doing, experiencing, and observing is usually very effective.

The desired computational tool should therefore provide us with not only robust numerical tools, but also easy ways to interpret and analyze the results of different scenarios. Mathcad is one of such tools that we will learn in this book.

Chapter 2
Getting Started with Mathcad

As in learning other software packages, in learning Mathcad, before we can develop more complicated procedures to solve mathematical problems, we need to understand Mathcad's basic features, such as definition and syntax. The effort needed for this purpose is minimal and much less than that needed for learning other software packages because Mathcad's basic features, as we will find later, are very intuitive. However, one should not overlook the details to avoid unnecessary mistakes. Keep in mind that bad input or instruction produces bad output or result (garbage in, garbage out).

2.1 Mathcad Worksheet and User Interface

If we are familiar with other program packages run on Windows, such as Microsoft Word or Excel, we will find the Mathcad user interface very familiar. As shown in Figure 2.1, Mathcad has Menu Bar, Standard Toolbar, Formatting Toolbar, and Math Toolbar. A tool bar that is different from that in many Windows applications is the Math Toolbar. This Math Toolbar makes mathematical calculations very easy and convenient to perform. The Math Toolbar allows a user to access all mathematical symbols and operations. If we click on any button on this bar we will see an expanded toolbar under the specified category on the screen. The following buttons are available from the Math toolbar:

1. Calculator Toolbar
2. Graph Toolbar
3. Vector and Matrix Toolbar
4. Evaluation Toolbar
5. Calculus Toolbar
6. Boolean Toolbar
7. Programming Toolbar
8. Greek Symbol Toolbar
9. Symbolic Keyword Toolbar

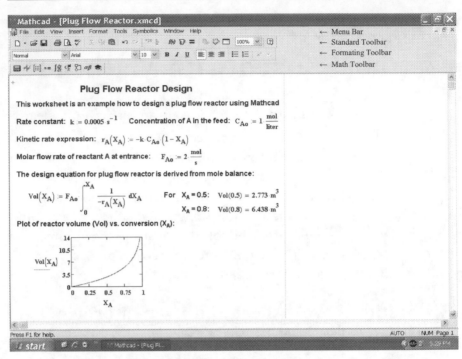

Figure 2.1. Mathcad Worksheet and User Interface

We can find out the name of each toolbar by letting the mouse pointer rest on a button. The function of each toolbar is pretty much self-explanatory. For example, the Calculator Toolbar contains all of the basic arithmetic operators we usually find in a calculator. Some of them can also be directly typed through the computer keyboard, for example "*" for multiplication, "/" for division, "^" for power, etc. We will learn and use most of the toolbars as needed in the subsequent sections and chapters.

Mathcad is essentially a worksheet for calculations. If we look at Figure 2.1 once again, one of the nicest features of Mathcad becomes obvious – we could develop a solution procedure just as we would do it on paper. We type the problem statement and variable definitions, then variable assignments, function definitions, and mathematical equations of the solution, and finally, graphical representation of the solution. Note that we can type math equations just as we would write them on paper. This is a natural and intuitive way to develop a problem solution, making it very easy to read and debug if necessary. Text, tables, equations, and graphs can be mixed on the same worksheet. To insert a text region, for example, type a double quotation mark (") (or click **Insert**, **Text Region** from the Menu Bar) and type the text we want. Thus, we can use Mathcad not only to obtain the results but also to elegantly present them.

2.2 Variables and Functions

The power and versatility of Mathcad become evident from the use of variables and functions. Mathcad has several built-in variables, such as π (3.14159), e (2.71828), and g (9.807 m/s^2), which can be directly used once we open a worksheet. However, most of the time we will use our own variables, which are called user-defined variables. A variable is defined by assigning a quantity to a variable name, for example:

$$x := 3$$

$$y := 2 \cdot x + 1$$

$$z := 1, 1.2 .. 5$$

Note that by default any assignment in Mathcad is done by typing a colon (":"), not an equal sign (when we type a colon, Mathcad will show a colon and an equal sign). This assignment is sometimes called the "define as equal to" operator. In fact, we can override the default operator display in the **Worksheet Options** in the **Tools** menu, but this is not recommended for beginners.

As we can see, a number, an expression of other previously defined variables, or a range of values might be assigned to a variable. In the next section, we will learn that a matrix/vector can also be assigned to a variable and in the last chapter, we will learn that a component can also be assigned to a variable. Thus, the type of a variable is directly determined by the type of the value assigned.

An equal sign ("=") is used when we want Mathcad to display the value of a defined variable or evaluate an expression (or a function) and display the result. From the example above, if we want to display the value of variable y, then we type "y" and "=":

$$y = 7$$

The following examples show the use of equal sign for evaluating an expression and a defined variable:

$$3 \cdot y - 1 = 20$$

$$w := x + 5 = 8$$

The variable z in the example is defined as a range variable, which has a range of values. In this assignment, the initial value is typed first, followed by a comma (","), and the second value, then a semi-colon (";"), and the last value. If the second value is omitted, the default interval is 1 (interval = the difference between the second and the first values).

Names in Mathcad can be a combination of alphanumeric, Greek letters, prime symbol, subscript (literal subscript), and underscore, but it should start with an alphabet or a Greek letter, which can be accessed from the Greek Toolbar. Mathcad distinguishes uppercase and lowercase letters. This name convention holds for any names including names for functions, matrices, and vectors.

Variable names may also contain operator symbols, as shown below

$$\left[\frac{P}{\rho \cdot R \cdot T} \right] := 1$$

This can be done by pressing [Ctrl][Shift]J to insert a pair of brackets. Thus, the term on the left side is in fact the name of the variable, which has been defined to have a value of 1.

Mathcad has a huge array of built-in functions readily available once we open a worksheet. Click on the *f(x)* icon on the Standard Tool-bar to see all of the built-in functions. We can also define our own functions (user-defined functions), for example:

$$f(x) := 3 \cdot x^2 + \frac{4}{x} - 5 + \sin(x)$$

$$g(x,y) := \begin{pmatrix} x & 1 \\ 2 & y^2 \end{pmatrix} \cdot \begin{pmatrix} y & x \\ 1 & 2 \cdot x \end{pmatrix}$$

$$h(x,f) := x + f(x)$$

$$p(M) := M^{-1} + \begin{pmatrix} 3 & 4 \\ 1 & 2 \end{pmatrix}$$

The variables in the bracket on the left side are called arguments. The choice of argument(s) for a function in a certain problem could be the key to problem solving. The argument(s) can be of any type, i.e., scalar, vector, matrix, or function. In the example above, f and g have scalar arguments; h has scalar and function arguments, while p has a matrix argument. Similar to variable assignment, the type of a function is also determined by the type of the assigned value resulted from the expression defining the function. Thus, f and h are scalar functions while g and p are matrix functions. See the next section for information on vector/matrix.

Mathcad always performs the calculations from left to right and from top to bottom, which is intuitive. This means that before a user-defined variable/function can be used for a certain mathematical operation, it should have been defined, i.e. given a value or an expression, somewhere above or on the left side of the operation. There is an exception for this general rule of the calculation flow, as described in the later section.

There are two other things that should be remembered when defining a variable/function. First, if the user-defined name is the same as one of the built-in names, then the user-defined variable/function will override the built-in variable/function. Second, variable names and function names in a worksheet are not distinguishable. It means that if we define a function y(x) and later we define a variable y, then we will not be able to use y(x) anywhere below the definition of the variable y. When we redefine a variable or a function, Mathcad will warn us by flagging it with a green wavy underscore. It is always a good idea to avoid this green wavy underscore by just changing the name of the variable/function.

2.3 Arrays: Vector and Matrix

Mathcad can also handle arrays. Thus, variables can also be defined as array variables, such as vectors and matrices. A vector is a matrix with a single row or column. The initialization (assignment) of an array can be done in three different ways:

(1) Using the Vector/Matrix icon from the Vector/Matrix Toolbar

The assignment is done by typing the name of the array and a colon, and then followed by choosing Matrix from the Matrix Toolbar or pressing [Ctrl-M]. Examples of this initialization are shown below:

$$A := \begin{pmatrix} 1 & 4 & 0 \\ 2 & 1 & 3 \\ -1 & 0 & 2 \end{pmatrix} \qquad B := \begin{pmatrix} 2 \\ 1 \\ 5 \end{pmatrix}$$

If the number of elements of an array is more than 600, the assignment must be done using other ways described below.

(2) Using a range variable as an array subscript (index variable)

This type of initialization is shown in the following examples:

$$i := 0..2 \qquad B_i := \qquad\qquad j := 0..4 \qquad C_{i,j} := i + j$$

$$\begin{array}{|c|} \hline 2 \\ \hline 1 \\ \hline 5 \\ \hline \end{array} \qquad B = \begin{pmatrix} 2 \\ 1 \\ 5 \end{pmatrix} \qquad C = \begin{pmatrix} 0 & 1 & 2 & 3 & 4 \\ 1 & 2 & 3 & 4 & 5 \\ 2 & 3 & 4 & 5 & 6 \end{pmatrix}$$

Here, i and j are range variables. The index of B, shown as a subscript, is typed using the left square bracket ("[") key followed by the name of the range variable, which is "i" in this case. After typing a colon, the values of elements of B, i.e., B_0, B_1, and B_2, are then entered. From one element to the next, we type a comma (",") and Mathcad automatically creates an empty placeholder. In this example, we define $B_0 = 2$, $B_1 = 1$, and $B_2 = 5$.

This initialization is handy when the elements of an array have a certain pattern, such as those of matrix C above, and the number of elements is large. The index of C is also typed using the left square bracket key followed by the name of the first range variable, a comma, and the name of the second range variable.

(3) Using data table

If we use data table, click **Insert, Data, Table** from the Menu Bar, and then type the name of the array. An example of this initialization is shown below:

A :=

	0	1	2
0	1	4	0
1	2	1	3
2	-1	0	2

Note that the default starting index in Mathcad is in fact 0, not 1. Thus, for example, the first element of A is

$$A_{0,0} = 1$$

To set the starting index of arrays to 1, click **Tools, Worksheet Options** in the Menu Bar and change the Array Origin to 1, or by redefining the built-in variable ORIGIN on the worksheet:

$$\text{ORIGIN} \equiv 1$$

where symbol "$\equiv$" means that the redefinition is globally applied throughout the worksheet (this symbol can be accessed in the Evaluation

Toolbar). This global assignment clearly overrides the rule of calculation process (from left to right and from top to bottom).

Please be aware that if we use a range variable to initiate an array and we still use the default starting index, the range variable must start from 0, otherwise the initialization will be incorrect, as shown below:

$$i := 1..3 \qquad B_i :=$$

$$\boxed{\begin{array}{c} 2 \\ 1 \\ 5 \end{array}} \qquad B = \begin{pmatrix} 0 \\ 2 \\ 1 \\ 5 \end{pmatrix}$$

which means we created a 4×1 vector (B_0 will automatically be set to 0 when it is not defined)!

In general, an index of an array can be a variable, a number, or an arithmetic expression. The following examples are legitimate indexed variables (again the subscripts are created using the left square bracket key):

$$B_2 \qquad\qquad C_{i+3} \qquad\qquad D_{2\cdot(3\cdot i-1)}$$

This index should not be confused with a literal subscript that is used as a part of the variable name. A literal subscript can be typed using a period (".") key. The following are examples of variables with literal subscript:

$$P_{sat} \qquad\qquad r_2$$

We cannot have an arithmetic expression for this subscript because literal subscript is not a variable/number. Note that r_2 here is a variable name, not an element of an array r.

Many array operations can be done easily in Mathcad, such as array addition and subtraction, dot product, cross product, matrix multiplication, matrix inversion and transposition, and calculation of determinant. For example:

$$A := \begin{pmatrix} 3 & 2 \\ 1 & 4 \end{pmatrix} \qquad B := \begin{pmatrix} 5 & 1 \\ 0 & 2 \end{pmatrix}$$

$$C := (A \cdot B)^{-1} = \begin{pmatrix} 0.09 & -0.07 \\ -0.05 & 0.15 \end{pmatrix} \qquad Det := |C| = 0.01$$

$$v := \begin{pmatrix} 1 \\ -2 \\ 3 \end{pmatrix} \qquad w := \begin{pmatrix} 4 \\ 1 \\ -1 \end{pmatrix} \qquad u := v \times w = \begin{pmatrix} -1 \\ 13 \\ 9 \end{pmatrix}$$

There is a unique type of array operator, i.e., vectorize operator, which is very handy for calculations involving a lot of data. This vectorize operator applies any arithmetic operator and/or function element-wise (element by element). To apply vectorize operator, we use an arrow sign above the operation or function, which can be accessed from the Matrix Toolbox. For example:

$$w := \begin{pmatrix} 4 \\ 3 \\ 2 \end{pmatrix} \qquad p := \overrightarrow{\left(\frac{\ln(w)}{w^2 + 1} \right)} = \begin{pmatrix} 0.082 \\ 0.11 \\ 0.139 \end{pmatrix}$$

$$f(x) := x^2 + 3 \qquad \overrightarrow{f(w)} = \begin{pmatrix} 19 \\ 12 \\ 7 \end{pmatrix}$$

Although many functions and operators, when applied to vectors, automatically perform operations on the elements of those vectors (arrays), the use of vectorize operator is recommended to avoid unnecessary mistakes. An example below shows a potential problem:

$$\overrightarrow{(w \cdot w)} = \begin{pmatrix} 16 \\ 9 \\ 4 \end{pmatrix} \qquad w \cdot w = 29$$

The second operation (without the vectorize operator) is in fact a dot product of two vectors.

2.4 Working with Units: Built-in and User-defined

In Mathcad, although there are some restrictions, physical units can be incorporated into most of calculations. Built-in and user-defined units can be considered as built-in and user-defined variables. When we want to incorporate a physical unit with a value, we just multiply the value with the name of the unit, such as shown in the example of area calculation below:

$$D := 2.5 \cdot in$$

$$A := \frac{\pi}{4} \cdot D^2$$

If the unit is a built-in unit, the name of the unit can be typed directly or inserted by clicking **Insert, Unit** from the Menu Bar. Knowing the exact names of the built-in units we will be using is important to avoid unnecessary error. One example of common mistake: the name of unit 'gram' in Mathcad is incorrectly thought as 'g', but the correct one is in fact 'gm'. If we forget about this, Mathcad will not warn us because 'g' is a built-in variable (gravitational acceleration). Therefore, use **Insert, Unit** from the Menu Bar if we are not sure about the unit names. If the unit is a user-defined unit, the unit should have been previously defined in term of built-in units and the name of the unit is also typed directly when it is used. The following is an example how we define and use a user-defined unit:

$$\text{kbar} := 1000 \cdot \text{bar} \qquad P := 2 \cdot \text{kbar}$$

Mathcad always stores values and displays all results in SI units (the default unit system is SI system). Changing the default unit system is possible but not recommended. For the example above, if we evaluate A, we get

$$A = 3.167 \times 10^{-3} \, m^2 \; \blacksquare$$

Note that an empty unit placeholder appeared on the right side (a small black solid box). We can easily change any unit displayed to the desired unit by typing the unit name in the empty unit placeholder. For example, if area in ft² is desired, we type "ft²" in the unit placeholder:

$$A = 0.034 \text{ft}^2$$

If another area unit is then desired, for example mm², just replace "ft²" with "mm²". There is no unit placeholder anymore once the unit placeholder has been filled.

Calculations that incorporate °C and °F as the temperature units can also be performed. For example:

$$t := 25\,°C = 298.15\,K \qquad t = 77\,°F \qquad t1 := 30\,°C$$

$$cp := 1 \cdot \frac{\text{cal}}{\text{gm} \cdot \Delta °C} \qquad \Delta H := cp \cdot (t1 - t) = 5 \frac{\text{cal}}{\text{gm}}$$

Note that there are two types of Celsius and Fahrenheit built-in units, i.e., temperature unit (°C or °F) and temperature difference unit (Δ°C or Δ°F), which should be distinguished from each other. These temperature units can be inserted by clicking **Insert, Unit** from the Menu Bar. The temperature unit °C or °F is in fact a postfix operator/function, for which, unlike the temperature difference units and other units, a multiplication operator ("dot") must not be used. If we use absolute temperature units, both temperature and temperature difference units are the same, i.e., K or R.

Remember that units are treated like variables. Consequently, if we define a variable named m, this will confuse ourselves with the unit "meter". Thus, if units are incorporated in our calculations, we need to avoid using unit names as the names of our user-defined variables/functions, particularly when the unit will be used. If this should happen, i.e., if a unit name needs to be used as a user defined variable/function, we need to insert a namespace operator to differentiate the unit name from the user-defined variable. For example:

$$m := 3 \qquad r := 0.5 \cdot m_{[unit]}$$

The namespace operator [unit] is created by pressing [Ctrl][Shift]N and typing the word "unit". The m in the definition of r above is the unit "meter", not the variable m, which has been defined to have a value of 3.

Since some Mathcad built-in functions do not support or fully support units, as we will find in the subsequent chapters, the unit incorporation in the calculations is in fact recommended for advanced users. For beginners, the units of all variables can be made consistent before calculations, which can be easily done in Mathcad, and then the calculations could be performed without incorporating any units.

2.5 Graphics Features: x-y Plot of Data and Functions

Mathcad has many graphic features, including 3-D plot, contour plot, and animation. In this section, we discuss only x-y plot because it is the most widely used graphic representation. We will learn other plots in later chapters as needed. Data and functions can be plotted easily in Mathcad after inputting the data or defining the functions. Data pairs should be inputted in the forms of vectors of the same size. For example, if we want to plot two quantities obtained from an experiment, where x is the independent variable and y is the dependent variable, we need to create vectors x and y first. A graph frame is created by clicking the Graph Toolbar from the Math Toolbar and choosing x-y plot (scattered plot). To let Mathcad know what we want to plot, we have to type the variable/function names in the empty central placeholders of the abscissa and ordinate (x- and y-axes). The other empty placeholders are for scale adjustment. Mathcad will automatically plot the data once we hit [enter] or click any region outside the graph region. If we want to plot a second experimental data z measured at the same set of independent variable x, we need to create the vector z before the graph region (remember that Mathcad processes any calculations from left to right and from top to bottom). To incorporate this second dependent variable in the existing graph, we click on the name of the first dependent variable on that graph and type a comma (",") immediately after the last letter of the name of the first dependent variable. Then we type "z" in the second placeholder that just appeared. If the second set of data is measured at a different set of independent variables, say x1, the vector x1 has to be created first along with the vector z and remember that x1 and z must be vectors of the same size. Besides inserting z at the y-axis on the graph, we also insert x1 at the x-axis by typing a comma (",") immediately after the last letter of the name of the first independent variable.

Below is an example of how to create x-y plot:

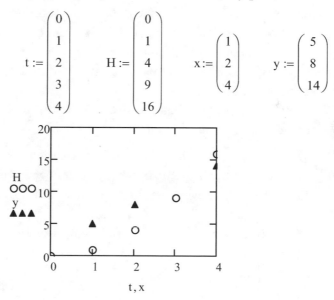

Of course, the vectors can also be created using other methods as described in Section 2.3.

Sometimes we need to format our graph to obtain a better presentation. For example, we can change the symbols from circles to squares or change the symbol plot to line plot. We can also type the plot title, axis labels, and legends. To format a graph, double click the graph and a new window for formatting pops up. The formatting features are straightforward and the readers are encouraged to try by themselves. If we need to know the values of the independent and dependent variables for a certain point in the graph, we can trace the plot by right clicking on the graph and choose **Trace** from the pop-up menu.

For plotting functions, it is a good practice to define the functions as described in section 2.2 before we create a graph frame, although we can directly type the expressions in the empty placeholder of the y-axis. To create the graph, again a graph frame is created by clicking the Graph Toolbar from the Math Toolbar and choosing x-y plot (scattered plot). For Quick Plot, we type the independent variable name on the x-axis, the function name on the y-axis placeholder, and hit [enter] or click any region outside the graph region. Mathcad plots the function at an independent variable range of [-10,10] by default. To plot the function at different independent variable range, the minimum and maximum values in the range must be typed in the empty placeholder of the x-axis. In plotting a function, in fact, Mathcad calculates the function at points of certain interval, which is, by default, a thousandth of the width of the independent variable range. To plot the function at a different interval, the independent variable must be defined as a range variable before the graph region. Other functions can also be plotted on the same graph by defining the functions before the graph region and inserting the function names at the y-axis (a comma after the first function name is also used to create an empty placeholder).

Recall that when we redefine a variable or a function, Mathcad will warn us by flagging it with a green wavy underscore. However, Mathcad will not warn us when we use a defined variable or built-in variable/unit as the independent variable in a plot because we in fact do not redefine that variable. The following example illustrates this potential problem:

$$A(R) := \pi \cdot R^2$$

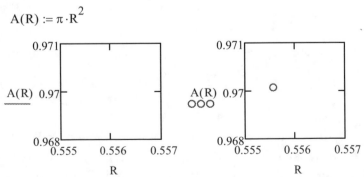

In this example, R is a built-in temperature unit (Rankine), which is equal to $(5/9) \cdot K$. Therefore Mathcad only plot one point, as shown in the right graph, which cannot be seen if the plot format is line plot (left graph). To plot the function A properly, we can use R1 instead of R for the independent variable or redefine R as a range variable.

Before we proceed to the next section, it is important to know how to control the units used in x-y plot when units are incorporated in our calculations. It has been mentioned in the previous section that Mathcad stores values and displays all results in SI units. This also applies to the graphical features. Mathcad displays x-y plots with x and y values in SI units. The following example illustrates the use of units in x-y plot:

$$A(r) := \pi \cdot r^2 \qquad r := 2 \cdot in, 2.1 \cdot in .. 8 \cdot in$$

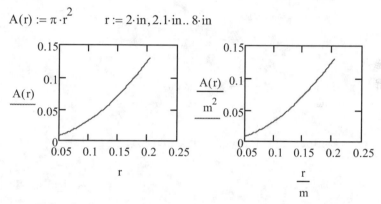

In the example above, we want to plot A(r) in the r range of 2 to 8 in. The independent variable r should be defined as a dimensional range variable, i.e., a range variable that has a unit. When we define a dimensional range variable, we must provide the second value to inform Mathcad about the step. Although both graphs represent the same plot, the right graph is the recommended plot because one will immediately understand that the units of r and A(r) are m and m^2, respectively. The units are entered as if they are divisors, for example, for the x-axis, we type [r][/][m]. To display the plot in different units, replace the units in the denominators with the desired units:

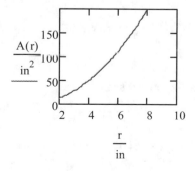

2.6 Symbolic Math Capabilities

Mathcad has some symbolic math capabilities. This means that in some cases, we can use Mathcad to analytically or symbolically solve math calculations. For example, in engineering we frequently need to perform differentiations, integrations, or matrix inversions analytically, which can be done easily in Mathcad (of course, if analytical solution does exist). This symbolic math could be done using a live symbolic equal sign, i.e., "→", or a symbolic keyword and a live symbolic equal sign. For example:

Evaluating a definite integral

$$\int_a^b \frac{x+1}{x^2+1}\,dx \to atan(b) - atan(a) - \frac{\ln(a^2+1)}{2} + \frac{\ln(b^2+1)}{2}$$

Evaluating the inverse of a matrix

$$\begin{pmatrix} x & -1 & 2 \\ 0 & 1 & 3 \\ 1 & 1 & 0 \end{pmatrix}^{-1} \to \begin{pmatrix} \dfrac{3}{3\cdot x+5} & -\dfrac{2}{3\cdot x+5} & \dfrac{5}{3\cdot x+5} \\ -\dfrac{3}{3\cdot x+5} & \dfrac{2}{3\cdot x+5} & \dfrac{3\cdot x}{3\cdot x+5} \\ \dfrac{1}{3\cdot x+5} & \dfrac{x+1}{3\cdot x+5} & -\dfrac{x}{3\cdot x+5} \end{pmatrix}$$

Simplifying an expression

$$\frac{x-1}{x^2-3\cdot x+2}\ \text{simplify}\ \to \frac{1}{x-2}$$

Collecting coefficients of a polynomial

$$P\cdot(v-b)\cdot v^2 - R\cdot T\cdot v^2 + a\cdot(v-b)\ \text{coeffs}, v\ \to \begin{pmatrix} -a\cdot b \\ a \\ -R\cdot T - P\cdot b \\ P \end{pmatrix}$$

Solving an equation symbolically

$$\left(\frac{P}{\rho\cdot R\cdot T} = 1 + B\cdot\rho\right)\ \text{solve}, \rho\ \to \begin{bmatrix} \dfrac{\sqrt{R^2\cdot T^2 + 4\cdot B\cdot P\cdot R\cdot T} - R\cdot T}{2\cdot B\cdot R\cdot T} \\ \dfrac{R\cdot T + \sqrt{R\cdot T\cdot(4\cdot B\cdot P + R\cdot T)}}{2\cdot B\cdot R\cdot T} \end{bmatrix}$$

These symbolic keywords and equal sign are available in the Symbolic Keyword Toolbar from the Math Toolbar. A symbolic keyword is needed to perform a certain operation or give an additional constraint/condition. In the examples above, the symbolic keyword "simplify" is used to simplify an arithmetic expression, "coeffs" is used to collect the coefficients of a polynomial in v, and "solve" is used to obtain the expression for ρ from the given equation (note that to type an equation, the equal sign is the bold equal sign, which can be obtained from the Evaluation Toolbar or by pressing [Ctrl][=]).

All examples in this chapter clearly demonstrate the synergy of power and elegance with which Mathcad performs its basic operations. There is no intention to cover all features or functions of Mathcad here, but rather to create an impression about what Mathcad can do and how easily math operations and plotting can be done. Various other important functions will be demonstrated in the following chapters and how to use them in solving chemical engineering problems is also presented.

Example set 2.1

1. Shell-and-tube heat exchangers are widely used in industries to exchange heat between hot and cold fluids. The effectiveness-NTU method is a calculation approach commonly applied for heat exchanger design and performance analysis.

   For 2-4 shell-and-tube heat exchanger:
   (a) Calculate the effectiveness (ε) for a heat capacity ratio (C_r) of 0.8 when the number of transfer units (NTU) is 2.
   (b) Create a plot of heat exchanger effectiveness (ε) as a function of the number of transfer units (NTU) in the range of $0 \le NTU \le 5$ for heat capacity ratios (C_r) of 0, 0.25, 0.5, 0.75, and 1.0.

   For a 2-4 shell-and-tube heat exchanger, the effectiveness can be obtained from:

   $$\varepsilon = \left[\left(\frac{1 - \varepsilon_1 \cdot C_r}{1 - \varepsilon_1} \right)^2 - 1 \right] \cdot \left[\left(\frac{1 - \varepsilon_1 \cdot C_r}{1 - \varepsilon_1} \right)^2 - C_r \right]^{-1}$$

   where ε_1 is the effectiveness for one shell of the heat exchanger given by

   $$\varepsilon_1 = 2 \cdot \left(1 + C_r + \sqrt{1 + C_r^2} \cdot \frac{1 + \exp\left(\frac{-NTU}{2} \cdot \sqrt{1 + C_r^2} \right)}{1 - \exp\left(\frac{-NTU}{2} \cdot \sqrt{1 + C_r^2} \right)} \right)^{-1}$$

   Solution:
   (a) We need to define C_r and NTU before we can calculate ε_1 and ε:

   $C_r := 0.8$ Note that r here is just a literal subscript, not an index.

   $NTU := 2$

Since ε is dependent on ε_1, calculate ε_1 first then ε:

$$\varepsilon_1 := 2 \cdot \left(1 + C_r + \sqrt{1 + C_r^2} \cdot \frac{1 + \exp\left(\dfrac{-NTU}{2} \cdot \sqrt{1 + C_r^2}\right)}{1 - \exp\left(\dfrac{-NTU}{2} \cdot \sqrt{1 + C_r^2}\right)}\right)^{-1} \qquad = 0.492$$

Note that 1 here is also a literal subscript, not an index.

$$\varepsilon := \left[\left(\frac{1 - \varepsilon_1 \cdot C_r}{1 - \varepsilon_1}\right)^2 - 1\right]\left[\left(\frac{1 - \varepsilon_1 \cdot C_r}{1 - \varepsilon_1}\right)^2 - C_r\right]^{-1} \qquad = 0.68$$

(b) To create a plot, we need to create the function ε, which is dependent on another function ε_1:

$$\varepsilon_1(C_r, NTU) := 2 \cdot \left(1 + C_r + \sqrt{1 + C_r^2} \cdot \frac{1 + \exp\left(\dfrac{-NTU}{2} \cdot \sqrt{1 + C_r^2}\right)}{1 - \exp\left(\dfrac{-NTU}{2} \cdot \sqrt{1 + C_r^2}\right)}\right)^{-1}$$

$$\varepsilon(C_r, NTU) := \left[\left(\frac{1 - \varepsilon_1(C_r, NTU) \cdot C_r}{1 - \varepsilon_1(C_r, NTU)}\right)^2 - 1\right]\left[\left(\frac{1 - \varepsilon_1(C_r, NTU) \cdot C_r}{1 - \varepsilon_1(C_r, NTU)}\right)^2 - C_r\right]^{-1}$$

Of course, we can use this function to solve part (a): $\qquad \varepsilon(0.8, 2) = 0.68$

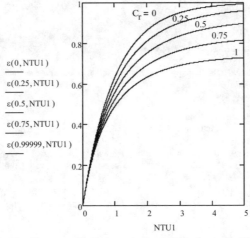

$\varepsilon(0, NTU1)$

$\varepsilon(0.25, NTU1)$

$\varepsilon(0.5, NTU1)$

$\varepsilon(0.75, NTU1)$

$\varepsilon(0.99999, NTU1)$

On the plot, the name of the independent variable is changed to NTU1 because NTU has been defined to have a value of 2 in part (a). If we used NTU on the plot, Mathcad would plot only one point at NTU = 2 for each C_r.

Note that for $C_r = 1$, the numerator and denominator of the equation of ε become zero. That is why on the plot we use $C_r = 0.99999$ for an approximation. We can also take the limit and assign the result to a different function:

$$\varepsilon_{Cr1}(NTU) := \lim_{C_r \to 1} \varepsilon(C_r, NTU) \to \frac{2\sqrt{2} - 2\sqrt{2} \cdot e^{-\sqrt{2} \cdot NTU}}{e^{-\sqrt{2} \cdot NTU} + 2 \cdot e^{-\dfrac{\sqrt{2} \cdot NTU}{2}} - 2\sqrt{2} \cdot e^{-\sqrt{2} \cdot NTU} + 2\sqrt{2} + 1}$$

In Mathcad, the limit operator, which can be accessed in the Calculus Toolbar, can only be used with the symbolic equal sign.

The function then can be used to plot or calculate the effectiveness for any NTU. For example, for NTU = 2:

$\varepsilon_{Cr1}(2) = 0.633 \qquad$ compared to $\qquad \varepsilon(0.99999, 2) = 0.633$

2. Air at 60°C (T_{inf}) flows over a long, 25-mm diameter cylinder with embedded electrical heater. In a series of test, measurements were made of the power per unit length, required to maintain the cylinder surface temperature at 300°C (T_s) for different free-stream velocity V of the air. From energy balance, this power per unit length is in fact the convection heat rate per unit length q'. The results are as follows:

Air velocity, V (m/s)	1	2	4	8	12
Heat rate, q' (W/m)	310	438	625	880	1080

Determine the convection coefficient (h) for each velocity and display your results graphically.

The convection heat rate per unit length is given by

$$q' = h \cdot A \cdot (T_s - T_{inf})$$

where A is the circumferential area of the cylinder per unit length.

Solution: We define all of the known variables and parameters:

$$T_{inf} := 60°C \qquad T_s := 300°C \qquad d := 25 \cdot mm$$

$$V := \begin{pmatrix} 1 \\ 2 \\ 4 \\ 8 \\ 12 \end{pmatrix} \cdot \frac{m}{s} \qquad q' := \begin{pmatrix} 310 \\ 438 \\ 625 \\ 880 \\ 1080 \end{pmatrix} \cdot \frac{W}{m} \qquad \text{The prime is typed by pressing [Ctrl][F7]}$$

The circumferential area of the cylinder per unit length needs to be calculated first:

$$A := \pi \cdot d = 0.079 m$$

The convection coefficient is calculated from the given equation:

$$h := \overrightarrow{\frac{q'}{A \cdot (T_s - T_{inf})}} = \begin{pmatrix} 16.446 \\ 23.237 \\ 33.157 \\ 46.685 \\ 57.296 \end{pmatrix} \cdot \frac{W}{m^2 \cdot K}$$

By using vectorize operator, all values of h at different velocities can be calculated at once.

Plot of h vs. V:

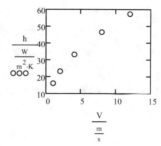

Problems

1. Explain the difference between the following function definitions:

$$h(x, f) := x + f(x)$$
$$h(x) := x + f(x)$$

2. Explain the difference between the following ORIGIN redefinitions:

$$ORIGIN \equiv 1$$
$$ORIGIN := 1$$

3. Implicit multiplication, i.e., the omission of "·" (multiplication operator), in mathematics is very common, e.g., $2x$, $3t$, etc. In fact, in
   Mathcad we are also not required to use this operator to type such expressions. However, this feature is not recommended. In other words,
   use the multiplication operator in any multiplications to avoid unnecessary mistakes. Explain the difference between the following expressions:

   $$h := y \cdot (x + 2)$$

   $$h := y(x + 2)$$

4. Fibonacci sequence is a sequence of numbers in which each number
   equals the sum of the two preceding numbers (of course, except the
   first two numbers, which are both equal to 1):

   $$1, 1, 2, 3, 5, 8, 13, \ldots$$

   Use Mathcad to enlist the first 30 numbers of this sequence. *Hint:*
   Create a range variable i from 3 to 30 and an array (indexed variable)
   f, where f_1 and f_2 are both set to 1. Then, create a recursive formula
   for f_i representing the i-th element in the Fibonacci sequence. Mathcad will automatically calculate the i-th element using the formula for
   all values of i. Note that this calculation is similar to "do loop" containing only one formula in FORTRAN or other computer languages.

5. A constant that is commonly used in chemical engineering is the gas
   constant R. The value of R can be obtained from Mathcad. Click **Help**,
   Reference Tables from the Menu Bar to open Mathcad Resources:
   Reference Tables window. Click on Fundamental Constants under
   Basic Science category and R can be found in Physico-Chemical Con-
   stants group and copied to an active Mathcad worksheet. The value is

   $$R := 8.314472 \frac{\text{joule}}{\text{mole} \cdot \text{K}}$$

   Using Mathcad, calculate R for the following units:
   a. $\text{atm} \cdot \text{cm}^3/(\text{mole} \cdot \text{K})$
   b. $\text{bar} \cdot \text{m}^3/(\text{mole} \cdot \text{K})$
   c. $\text{torr} \cdot \text{m}^3/(\text{kmole} \cdot \text{K})$. *Hint:* create a user-defined unit: kmole.

6. Second virial coefficient (B) is an important gas property. In fact, this
   property is related to the molecular interaction between two mole-
   cules. If we know the second virial coefficient of a gas, for a given
   pressure (P) and temperature (T) we can estimate the molar density
   (ρ) of that gas using the virial equation of state:

$$\frac{P}{\rho RT} = 1 + B\rho$$

where R is the gas constant. Note that the second term on the right side can be considered as a correction term, because for an ideal gas: $P/\rho RT = 1$; an ideal gas is a gas that does not have any molecular interactions.

The second virial coefficient can be estimated from statistical thermodynamics, for example for a square-well fluid, which is a model for more realistic fluid, it is given by

$$B = b_0 \lambda^3 \left(1 - \frac{\lambda^3 - 1}{\lambda^3} \exp \frac{\varepsilon}{kT} \right)$$

where $b_0 = \dfrac{2}{3} \pi N_A \sigma^3$ = the second virial coefficient of a hard-sphere fluid

σ = the diameter of the molecule

ε = the well depth of the molecular potential

λ = a dimensionless parameter related to the well width of the molecular potential

k = the Boltzmann constant

N_A = the Avogadro number

a. Plot the second virial coefficient of CO_2 as a function of temperature, say from 200 to 1200 K. The square-well parameters for CO_2 are $\sigma = 3.917$ Å, $\lambda = 1.83$, and $\varepsilon/k = 119.0$ K.

b. Plot on the same graph the second virial coefficient of CH_4 as a function of temperature. The square-well parameters for CH_4 are $\sigma = 3.40$ Å, $\lambda = 1.85$, $\varepsilon/k = 88.8$ K.

c. Calculate the molar density (in mol/cm^3) of CO_2 at $T = 680$ K and $P = 40$ bar using this virial equation of state.

d. To investigate the behavior of a real gas, plot ρ vs. T ($400 \leq T \leq 700$ K) at $P = 60$ bar and ρ vs. P ($0 \leq P \leq 60$ bar) at $T = 400$ K for CO_2 using this virial equation of state. On those graphs, also plot the molar density of an ideal gas. By analyzing these graphs, at what conditions would a real gas behave like an ideal gas? Do you know why? Write your comments.

7. A first order reaction is carried out in a series of N equal-size mixed flow reactors. The total conversion (X) achieved for this system is given by

$$\frac{1}{1-X} = (1+k\tau)^N$$

where k is the reaction constant (= 0.075 hr^{-1}) and τ is the mean residence time in a single mixed reactor.

a.   Plot the total conversion of this reaction as a function of τ ($0 \le \tau \le$ 40 hrs) for $N = 4$.

b.   Plot on the same graph the total conversion as a function of τ for $N = 2$.

c.   Calculate the total conversion if the number of equal-size mixed reactors is 6 and the residence time of the fluid in a single reactor is 5 hrs.

d.   Suppose we allocate a total of 10 hrs for this reaction ($\sum_i \tau_i = 10$ hrs, the summation is over all reactors) and a large conversion is desired, would you suggest having a lot of small mixed reactors or a few large mixed reactors? Explain your reasoning using some calculations. *Hint*: a larger reactor has a larger fluid residence time.

8. In chemical industry, the cost of piping system (pipes and fittings) and pumping are important costs that should be considered. The annual cost of a pipeline with a standard carbon steel pipe and a motor-driven centrifugal pump is given by:

$$C = 0.45L + 0.245LD^{1.5} + 325\sqrt{P} + 61.6P^{0.925} + 102$$

where L is the length of the pipeline in ft, D is the pipe diameter in inch, and P is the power of the pump in hp, which can be calculated from:

$$P = 4.4 \times 10^{-8} \frac{LQ^3}{D^5} + 1.92 \times 10^{-9} \frac{LQ^{2.68}}{D^{4.68}}$$

In the above equation, Q is the volumetric flow rate of the fluid in gpm.

a.   Plot the annual cost as a function of pipe diameter, say from 0.25 to 6 in, for 1000 ft pipeline with a fluid rate of 20 gpm.

b.   What is the annual cost needed if a pipe with a diameter of 1 in is used? Is there an optimum diameter for this pipeline? Explain why it is expensive when a very small pipe diameter is used.

9. In petroleum industries, it is common that the molecular weight of an oil fraction is unknown and thus must be estimated by using a correlation. A correlation that is widely used is developed by Riazi and Daubert, which is also referred to as the API (American Petroleum Institute) method. The correlation can be applied to hydrocarbons with molecular weight ranging from 70-700, which is nearly equivalent to boiling point range of 300-850 K, and the API gravity range of 14.4-93. This molecular weight (M) correlation is given as a function of the average boiling point (T_b) and the specific gravity at 60°F (SG) of the oil fraction of interest as follows:

$$M = 42.965 \exp(2.097 \times 10^{-4} T_b - 7.78712 SG + 2.08476 \times 10^{-3} T_b SG) T_b^{1.26007} SG^{4.98308}$$

where T_b is in K and SG is related to the API gravity (API):

$$SG = \frac{141.5}{API + 131.5}$$

a. Estimate the molecular weight of an oil fraction that has an average boiling point of 344.7°C and an API gravity of 50.

b. Plot on the same graph M vs. T_b ($400 \le T_b \le 600$ K) with API gravity = 20, 40, and 60.

Chapter 3
Nonlinear Equation

Nonlinear equations, both polynomial and non-polynomial, are commonly encountered in chemical engineering applications. In solving a nonlinear equation, it is well known that we may have more than one root. For a polynomial equation of order *n*, there are exactly *n* roots that satisfy the equation. If complex roots do exist, they must appear in pairs, i.e., complex numbers and their conjugates. For a non-polynomial equation, it is uncertain whether we have only one root or more. In many cases, we are only interested in real roots and most of the time only one real root has physical meaning, which is the solution of that particular physical condition. However, obtaining two or more real roots that have physical meaning is not uncommon, such as found in the phase equilibrium calculations. In chemical engineering, complex roots are important for process control analysis. Mathcad provides us with easy and simple procedures to solve a nonlinear equation.

In this chapter, we also introduce a procedure, referred to as parametric procedure, which is very useful in solving many parametric problems in chemical engineering. In a parametric problem, the unknown to be calculated is dependent on some parameters, the values of which will be varied. This parametric problem often occurs in what-if analysis.

3.1 Polynomial

Mathcad has two built-in functions that can be used to find roots of a polynomial, i.e., *polyroots* and *root*. In fact, *find* function with a *Given* block can also be used to find roots of polynomials or other nonlinear equations, but it will be discussed in the next chapter. The procedures of using *polyroots* and *root* functions are described in Procedures 3.1 and 3.2, respectively.

As can be concluded from Procedure 3.1, if the variable x of a polynomial f(x) has a unit, each coefficient of the polynomial will have different units. Thus, each element of the vector c should have different unit, which is not allowed in Mathcad. Therefore, *polyroots* function does not support units.

Procedure 3.1 : _polyroots_

The _polyroots_ function is the best function for finding all roots without providing any initial guesses.

For example, we want to find the roots of a cubic polynomial:

$$f(x) := 2 + 3 \cdot x - 5 \cdot (x - 1)^2 + x^3$$

In other words, we want to solve $f(x) = 0$.

1. Collect all coefficients of the polynomial using the symbolic math for collecting coefficients and assign them to an array variable (vector):

$$c := f(x) \ \text{coeffs}, x \ \rightarrow \begin{pmatrix} -3 \\ 13 \\ -5 \\ 1 \end{pmatrix}$$

Remember that when we do this, the variable x must have never been defined (no value has been assigned to x), otherwise f(x) is not considered as a polynomial, but a number.

2. Use _polyroots_ to find all roots (we must have three roots for a cubic polynomial):

$$x := \text{polyroots}(c) = \begin{pmatrix} 0.254 \\ 2.373 - 2.482i \\ 2.373 + 2.482i \end{pmatrix}$$

$$x_0 = 0.254 \qquad x_1 = 2.373 - 2.482i \qquad x_2 = 2.373 + 2.482i$$

In this case, we have one real root and two complex roots.

Parametric procedure for _polyroots_

For example, we want to find the roots of a cubic polynomial with one parameter a:

$$f(a, y) := a + 3 \cdot y - 5 \cdot (y - 1)^2 + y^3$$

Now, the parameter a must be included as an argument in all functions and dependent variables in the procedure that are dependent on the parameter.

$$c(a) := f(a, y) \ \text{coeffs}, y \ \rightarrow \begin{pmatrix} a - 5 \\ 13 \\ -5 \\ 1 \end{pmatrix}$$

$$y(a) := \text{polyroots}(c(a))$$

For a = 2:

$$y(2) = \begin{pmatrix} 0.254 \\ 2.373 - 2.482i \\ 2.373 + 2.482i \end{pmatrix}$$

For a = -15

$$y(-15) = \begin{pmatrix} 1.052 - 2.408i \\ 1.052 + 2.408i \\ 2.896 \end{pmatrix}$$

The roots for any a can be calculated easily.

As noted in Procedure 3.2, _root_ function requires an initial guess to find the root. In fact, the success of using this function depends on how close the provided initial guess to the real root is. If the initial guess is not close enough to the desired root, the _root_ function may give another root or fail to converge. If the _root_ function fails to converge, it will be colored in red and an error message will be shown when one clicks on it. Plotting the function that is being solved is therefore very important to obtain a good initial guess for a real root. Unlike _polyroots_ function, _root_ function does support units.

Procedure 3.2: _root_

root function can only find one root at a time. Although _root_ function can also find imaginary roots by using complex number as initial guess, _root_ function is mainly used to find the real roots of a polynomial.

For example, we want to find the roots of a cubic polynomial:

$$f(x) := x^3 + 5 \cdot x^2 + x - 3$$

In other words, we want to solve $f(x) = 0$.

1. Plot the function to get the idea about the number of real roots and the initial guess.

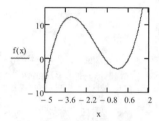

By analyzing this plot, we know that there are three real roots (three values of x) that make f(x) = 0.

2. Define an initial guess: $x := 1$

3. Use the _root_ function to find one of the roots that is close to the initial guess

$$x_1 := root(f(x), x) = 0.646$$

4. Repeat step 2 and 3 for other roots

Initial guess: $x := -2$

$$x_2 := root(f(x), x) = -1$$

Initial guess: $x := -5$

$$x_3 := root(f(x), x) = -4.646$$

The parametric procedure for _root_ function or any other function is similar to that for _polyroots_ function, where the parameters should be included as arguments in all functions and dependent variables that are dependent on the parameters. See Example 2 of Example set 3.1 in this chapter as an example.

Example set 3.1

1. Calculate the molar volume of carbon dioxide at 400 K and 50 bar using Peng-Robinson equation of state:

$$P = \frac{R \cdot T}{(V - b)} - \frac{a}{V^2 + 2 \cdot b \cdot V - b^2}$$

where R is the gas constant , T is the absolute temperature, V is the molar volume, and a and b are the parameters calculated from:

$$a = \frac{0.457235 R^2 \cdot Tc^2}{Pc} \cdot \left[1 + \left(0.37464 + 1.54226\omega - 0.26992\omega^2 \right) \cdot \left[1 - \left(\frac{T}{Tc} \right)^{0.5} \right] \right]^2$$

$$b = 0.077796 R \cdot \frac{Tc}{Pc}$$

where Tc is the critical temperature, Pc is the critical pressure, and ω is the accentric factor.

Data for carbon dioxide: Tc = 304.2 K, Pc = 73.83 bar, ω = 0.224

Solution:

As we can see from the equation above, to calculate the molar volume V, we are dealing with a polynomial. Since _polyroot_ will be used here, we cannot incorporate units in our calculations and thus the unit of the gas constant from the Reference Table should be made consistent first.

From the Mathcad Reference Table: $R := 8.314472 \dfrac{\text{joule}}{\text{mole·K}} = 83.145 \dfrac{\text{bar·cm}^3}{\text{mole·K}}$

Parameters and constants

     $R := 83.145$ $T := 400$ $P := 50$

     For CO_2: $Tc := 304.2$ $Pc := 73.83$ $\omega := 0.224$

Calculate parameters a and b

$$a := \frac{0.457235 R^2 \cdot Tc^2}{Pc} \cdot \left[1 + \left(0.37464 + 1.54226\omega - 0.26992\omega^2 \right) \cdot \left[1 - \left(\frac{T}{Tc} \right)^{0.5} \right] \right]^2$$

$$b := 0.077796 R \cdot \frac{Tc}{Pc}$$

Collect the coefficients of the polynomial we want to solve

$$f(V) := P \cdot (V - b) \cdot \left(V^2 + 2 \cdot b \cdot V - b^2 \right) - R \cdot T \cdot \left(V^2 + 2 \cdot b \cdot V - b^2 \right) + a \cdot (V - b)$$

$c := f(V) \begin{vmatrix} \text{coeffs}, V \\ \text{float}, 10 \end{vmatrix} \rightarrow \begin{pmatrix} -6.026418195e7 \\ 1.303801515e6 \\ -31925.43086 \\ 50.0 \end{pmatrix}$ The symbolic keyword 'float' is used to limit the number of significant figures in a result. In this case, we limit the number of significant figures to 10 for defining the coefficients.

Volume calculation using *polyroots*

$$V := \text{polyroots}(c) = \begin{pmatrix} 20.108 + 40.127i \\ 20.108 - 40.127i \\ 598.292 \end{pmatrix}$$

$\text{Volume} := V_2 = 598.292$ cm3/mol

2. A brick wall of 0.2 m thick (L) separates a hot combustion gas of a furnace from the ambient air and its surroundings. Under steady-state condition, the temperature of the inner surface of the brick wall (T_i) is 623.15 K and of the ambient air (T_a) is 293.15 K. Free convection heat transfer to the air is characterized by a convection coefficient (h) of 15 W/m^2.K. Thermal conductivity of the brick wall (k) is 1.2 W/m.K and its surface emissivity (ε) of 0.7. The temperature of the surrounding (T_s) can also be assumed 293.15 K.

a. What is the brick outer surface temperature (T_o)?

b. Plot the brick outer surface temperature (T_o) as a function of the brick inner surface temperature (T_i) asumming all of the other parameters constant $(400 \le T_i \le 800\text{K})$.

From energy balance:

Energy transfered to the outer surface = Energy transfered from the outer surface

$$k \cdot \frac{T_i - T_o}{L} = h \cdot (T_o - T_a) + \varepsilon \cdot \sigma \cdot \left(T_o^4 - T_s^4 \right)$$

where σ = Boltzmann constant = 5.67 10^{-8} W/m^2.K^4

Solution:

a. $T_i := 623.15\text{K}$ $T_a := 293.15\text{K}$ $T_s := 293.15\text{K}$ $L := 0.2 \cdot \text{m}$

$k := 1.2 \dfrac{W}{m \cdot K}$ $\sigma := 5.67 \cdot 10^{-8} \cdot \dfrac{W}{m^2 \cdot K^4}$ $h := 15 \cdot \dfrac{W}{m^2 \cdot K}$ $\varepsilon := 0.7$

$f(T_o) := h \cdot (T_o - T_a) + \varepsilon \cdot \sigma \cdot \left(T_o^4 - T_s^4 \right) - k \cdot \dfrac{T_i - T_o}{L}$ $T_o := 300\text{K}, 301\text{K} .. 600\text{K}$

Initial guess: $T := 400\text{K}$

           $T_o := \text{root}(f(T), T) = 367.078\text{K}$

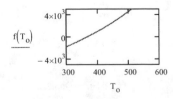

b. We can repeat the calculation in part (a) above several times for different T_i and plot the results manually. However, parametric procedure is the best way to do this, i.e., by making T_o as a function of T_i.

Include T_i in the function f: $f(T_i, T_o) := h \cdot (T_o - T_a) + \varepsilon \cdot \sigma \cdot \left(T_o^4 - T_s^4 \right) - k \cdot \dfrac{T_i - T_o}{L}$

$$T_o(T_i) := root\big(f(T_i, T), T \big)$$ Note that for the initial guess, we use the value of T in part (a).

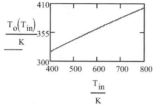

In this case, the value of T_o depends on parameter T_i. For every T_i, the corresponding value of T_o is calculated by using *root* function with the specified initial guess. If the range of T_i is very wide, of course there will be a possibility that *root* function gives other real root or fails to give any root. Thus, a different initial guess might be needed. Such a procedure is very important and should be clearly understood.

3. The reduction of iron ore (Fe_3O_4, molecular weight M_{ore} = 232 g/mole) of density ρ_{ore} = 4.6 g/cm^3 and size R = 5 mm by hydrogen (A) can be approximated by the unreacted core model. The reaction is:

$$Fe_3O_{4(solid)} + 4\,H_{2(gas)} \rightarrow 4\,H_2O_{(gas)} + 3\,Fe_{(solid)}$$

In this model, the reaction occurs first at the outer skin of the particle. The reaction boundary then moves into the inner particle as the unreacted solid Fe_3O_4 shrinks and leaves behind the inert solid product Fe, as depicted in the figure below. The radius of the particle can be assumed constant (R).

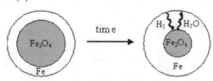

The hydrogen gas then must diffuse through the inert solid product and react when it reaches the unreacted core surface. The effective diffusivity of hydrogen through the product layer (D_e) = 0.03 cm^2/s and the concentration of hydrogen at the outer surface of the particle (C_{A1}) = 1.16 10^{-5} mole/cm^3. At the conditions of interest, it is known that mass diffusion controls the process, not the reaction kinetics. It means that the reaction is fast and the concentration of hydrogen at the surface of the unreacted core (C_{A2}) can be assumed zero.

If the average time that the particle spends in the reduction zone (t) is 40 min. What is the average size of the unreacted particle core (r_c) exiting the zone?

From material balance, the following expression can be derived:

$$t = \frac{2 \cdot \rho_{ore} \cdot R^2}{3 \cdot M_{ore} \cdot D_e \cdot (C_{A1} - C_{A2})} \left[1 - 3 \cdot \left(\frac{r_c}{R} \right)^2 + 2 \cdot \left(\frac{r_c}{R} \right)^3 \right]$$

Solution: $\rho_{ore} := 4.6$ $R := 0.5$ $D_e := 0.03$ $C_{A1} := 1.16\,10^{-5}$ $C_{A2} := 0$ $M_{ore} := 232$

$t := 40 \cdot 60$

The function to be solved: $f(r) := \dfrac{2 \cdot \rho_{ore} \cdot R^2}{3 \cdot M_{ore} \cdot D_e \cdot (C_{A1} - C_{A2})} \left[1 - 3 \cdot \left(\dfrac{r}{R} \right)^2 + 2 \cdot \left(\dfrac{r}{R} \right)^3 \right] - t$

$c := f(r) \begin{vmatrix} coeffs, r \\ float, 10 \end{vmatrix} \rightarrow \begin{pmatrix} 7095.970406 \\ 0 \\ -113951.6449 \\ 151935.5265 \end{pmatrix}$ $r_c := polyroots\,(c) = \begin{pmatrix} -0.219 \\ 0.336 \\ 0.634 \end{pmatrix}$

Then the size of the unreacted particle core is 3.36 mm (one root is negative and the other is greater than the initial size of the particle).

3.2 Non-Polynomial

For finding roots of a non-polynomial equation, *root* function will do the job. As in using this function for polynomial problems, the initial guess is very important. If the initial guess is not close enough to the desired root, the *root* function may give another root or fail to converge.

Example set 3.2

1. An electric wire with a bare radius (r_1) of 1 mm and a surface temperature (T_1) of 400 K is covered with an insulation. The temperature of the surrounding air (T_o) is 298 K. The convective heat coefficient (h_o) is 20 W/(m^{2}K) and the conductivity of the insulation (k) is 0.4 W/m.K.

   a. Plot the heat loss per meter of wire as a function of the radius of the insulated wire (r_2), say from 1 mm to 60 mm, and discuss whether the heat loss decreases as we add more insulation. Assume that T_1 is constant and not affected by the insulation thickness.

   b. Calculate the thickness of the insulation if the heat loss per meter is 60 W.

The heat loss can be calculated from
$$q = \frac{2 \cdot \pi \cdot L \cdot \left(T_1 - T_o\right)}{\dfrac{\ln\left(\dfrac{r_2}{r_1}\right)}{k} + \dfrac{1}{r_2 \cdot h_o}}$$

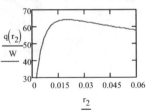

where   L = the length of the wire

Solution: $L := 1 \cdot m$ $r_1 := 1 \cdot mm$ $k := 0.4 \cdot \dfrac{W}{m \cdot K}$ $h_o := 20 \cdot \dfrac{W}{m^2 \cdot K}$ $T_1 := 400 \cdot K$ $T_o := 298 \cdot K$

a. Create a function for heat loss:

$r_2 := 1 \cdot mm, 2 \cdot mm .. 60 \cdot mm$

$$q\left(r_2\right) := \frac{2 \cdot \pi \cdot L \cdot \left(T_1 - T_o\right)}{\dfrac{\ln\left(\dfrac{r_2}{r_1}\right)}{k} + \dfrac{1}{r_2 \cdot h_o}}$$

[Plot: vertical axis $\dfrac{q\left(r_2\right)}{W}$ ranging 30 to 70; horizontal axis $\dfrac{r_2}{m}$ ranging 0 to 0.06, with tick marks 0.015, 0.03, 0.045]

At first, by adding insulation, the heat loss increases! After r_2 reaches a critical value at 20 mm, adding more insulation results in the decrease of heat loss.

b. $f\left(r_2\right) := q\left(r_2\right) - 60 \cdot W$ There are 2 roots for f(r_2) = 0, as can be analyzed from the graph above.

   Initial guess: $r_2 := 15 \cdot mm$ Insulation thickness:

   $r_2 := \text{root}\left(f\left(r_2\right), r_2\right) = 10.315 \text{mm}$ $t := r_2 - r_1 = 9.315 \text{mm}$

   Initial guess: $r_2 := 50 \cdot mm$

   $r_2 := \text{root}\left(f\left(r_2\right), r_2\right) = 46.745 \text{mm}$ $t := r_2 - r_1 = 45.745 \text{mm}$

2. Calculate the bubble point temperature (t_{bubble}) of a mixture containing 0.3 mol fraction benzene (x_B) and 0.7 mol fraction toluene (x_T) at a pressure (P) of 1 atm. The vapor pressure (P_{sat}) of these compounds as a function of temperature can be represented well by Antoine equation:

   Antoine equation: $Psat_i(t) = \exp\left(A_i - \dfrac{B_i}{t + C_i}\right)$ mmHg

   where A_i, B_i, and C_i are Antoine parameters for component i and t is the temperature [^{o}C]

The bubble point temperature can be calculated by solving the following equation:

$$P = x_B \cdot Psat_B(t_{bubble}) + x_T \cdot Psat_T(t_{bubble})$$

Data: Benzene: A = 15.90085, B = 2788.507, C = 220.790

Toluene: A = 16.01066, B = 3094.543, C = 219.377

Solution:

Benzene: $A_B := 15.90085$ $B_B := 2788.507$ $C_B := 220.790$ $x_B := 0.3$

Toluene: $A_T := 16.01066$ $B_T := 3094.543$ $C_T := 219.377$ $x_T := 1 - x_B$

$P := 760$ (1 atm = 760 mmHg)

Antoine equation: $Psat(t, A, B, C) := \exp\left(A - \dfrac{B}{t + C}\right)$

Create a function for bubble point calculation:

$$f(t) := x_B \cdot Psat(t, A_B, B_B, C_B) + x_T \cdot Psat(t, A_T, B_T, C_T) - P$$

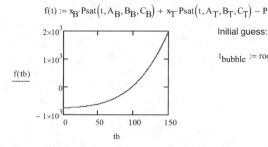

Initial guess: $t := 100$

$t_{bubble} := root(f(t), t) = 98.456$ °C

3. A heat exchanger is needed in a chemical process. One possible choice would be a standard type, which costs $4000. This heat exchanger will have a useful life of 6.5 years and a scrap value of $300. Another possible choice of equivalent design capacity costs $7000 but will have a useful life of 10 years and a scrap value of $800.

a. Assuming an effective compound interest rate of 6% (i = 0.06) per year, determine which heat exchanger gives the least yearly cost due to depreciation.

b. At what effective compound interest rate the two types of heat exchanger give the same yearly depreciation cost?

Yearly cost due to depreciation can be calculated from: $D = (C - S) \cdot \dfrac{i}{(1 + i)^n - 1}$

where C is the initial cost of the equipment, S is the scrap value of the equipment at the end of its useful life, n is the life time of the equipment in years, and i is the compound interest rate per year.

Solution: Create the depreciation function: $D(C, S, n, i) := (C - S) \cdot \dfrac{i}{(1 + i)^n - 1}$

a. Compound interest rate: $i := 0.06$

Heat exchanger I: $C_1 := 4000$ $S_1 := 300$ $n_1 := 6.5$

Yearly cost due to depreciation: $D(C_1, S_1, n_1, i) = 482.132$

Heat exchanger II: $C_2 := 7000$ $S_2 := 800$ $n_2 := 10$

Yearly cost due to depreciation: $D(C_2, S_2, n_2, i) = 470.381$ (the least)

b. Construct a function: $f(i) := D(C_1, S_1, n_1, i) - D(C_2, S_2, n_2, i)$

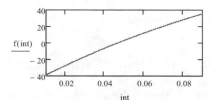

Initial guess: $i := 0.05$

$i := root(f(i), i) = 0.047$

The compound interest rate is 4.7%

Problems

1. Plot the molar volume of CO_2 as a function of temperature from 400 to 700 K at 50 bar using Peng-Robinson equation of state as described in Example 1 of Example set 3.1. *Hint:* Treat the temperature as a parameter.

2. Repeat the molar volume calculation for CO_2 at 283 K and 50 bar using Peng-Robinson equation of state as described in Example 1 of Example set 3.1. At this condition, it is known that CO_2 is a liquid. As you will also find, at this condition, the equation of state will give you three real positive roots. When a cubic equation of state gives three positive roots, only the smallest and/or the largest roots have physical meaning. We should never use the other root in volume calculations from a cubic equation of state because this root is unphysical. Use your logical thinking to determine which of the roots is the molar volume of *liquid*

3. A stirred-tank reactor processes a reaction mixture at isothermal condition (50°C). The reaction is

$$A + 2B \rightarrow C$$

and the kinetic rate expression is given by: $-r_A = \dfrac{k_1 C_A C_B^2}{1 + k_2 C_A}$

where k_1 = the first reaction constant = 0.01 $(kgmol/m^3)^{-2}(hr)^{-1}$

k_2 = the second reaction constant = 0.5 $(kgmol/m^3)^{-1}$

C_A = concentration of A in the product stream $(kgmol/m^3)$ = F_A/v

C_B = concentration of B in the product stream $(kgmol/m^3)$ = F_B/v

v = volumetric flow rate of the product stream = 2 $m^3/hour$ (assume that the volumetric flow rate of the product stream does not change appreciably with the mixture composition, i.e., v is constant).

F_A = molar flow rate of A in the product stream (kgmol/hour)
= $F_{A0}(1 - X_A)$

F_B = molar flow rate of B in the product stream (kgmol/hour)
= $F_{B0} - 2F_{A0}X_A$

F_{A0} = molar flow rate of A in the feed stream = 9 kgmol/hour

F_{B0} = molar flow rate of B in the feed stream = 18 kgmol/hour

X_A = conversion of A = $\dfrac{\text{moles of A reacted}}{\text{moles of A fed}}$

The design equation for a stirred-tank reactor is obtained from mass balance:

$$V = \frac{F_{A0}X_A}{-r_A}$$

where V = volume of reactor (m³)

a. Calculate the conversion of A if V = 50 m³.

b. Plot the conversion of A as a function of volume of reactor ($0 \le V \le$ 70 m³).

4. The virial equation of state that relates pressure (P), temperature (T), and density (ρ) is:

$$\frac{P}{\rho RT} = 1 + B\rho + C\rho^2 + D\rho^3$$

where R = gas constant, and B, C, and D are the second, third, and fourth virial coefficients, respectively.

For nitrogen at 200 K: B = −0.0361 liter/mol, C = 2.7047×10⁻³ (liter/mol)², and D = −4.4944×10⁻⁴ (liter/mol)³.

a. Determine the density of nitrogen (in mol/liter) at 200 K and 9 atm.

b. Plot the density of nitrogen at 200 K as a function of pressure ($1 \le P \le 30$ atm).

Note: At 200 K and pressure up to 33 atm, it is known that the density of nitrogen is always less than its critical density (ρ_c = 11.1839 mol/liter).

5. Production of propylene glycol by hydrolysis of propylene oxide is carried out in an adiabatic continuous-stirred tank reactor (CSTR) [Fogler, *Elements of Chemical Reaction Engineering*, Prentice-Hall, Englewood Cliffs, 1992]. The reaction can takes place at low temperature when sulfuric acid is used as catalyst:

propylene oxide (O) + water (W) → propylene glycol

The standard heat of reaction (ΔH_R^0 at T_R = 68°F) for this reaction is −36400 Btu/lbmol propylene oxide. The feed containing 36 mol%

propylene oxide and 64 mol% methanol enters the reactor at 75°F (T_{in}).

a. Determine the temperature of the product stream (T) if the reactor volume (V) is 350 gallons.
b. Determine the conversion (X).

The following equations can be obtained from material and energy balances:

Material balance:
$$X = \frac{(16.96 \times 10^{12} \text{ hr}^{-1})e^{-\frac{E}{R_g T}}\frac{V}{v_0}}{1 + (16.96 \times 10^{12} \text{ hr}^{-1})e^{-\frac{E}{R_g T}}\frac{V}{v_0}}$$

Energy balance:
$$X = \frac{\widetilde{C}_{p,in}(T - T_{in})}{-\left\{\Delta H_R^0 + \Delta C_p(T - T_R)\right\}}$$

where X = conversion
v_0 = volumetric flow rate entering the reactor = 326.0 ft³/hr
E = activation energy = 32400 Btu/(lbmol)
R_g = gas constant = 1.987 Btu/(lbmol·R)
$\widetilde{C}_{p,in}$ = heat capacity of the mixture entering the reactor per lbmol of propylene oxide = 400.0 Btu/(lbmol·R)
ΔC_p = the overall change in heat capacity per lb mol of propylene oxide reacted = –7 Btu/(lbmol·R)

6. In a tempering process, glass plate, the temperature of which is initially uniform (T_i = 300°C), is cooled for 10 minutes (t) using airflow over both sides of the plate, for which its temperature (T_∞ = 30°C) and convection heat transfer coefficient (h = 100 W/(m²·K)) are kept constant. If the thickness of the plate ($2L$) is 30 mm and radiation exchange is neglected, what is the midplane temperature (T_0) at the end of the cooling process? Data for glass: thermal conductivity (k) = 1.2 W/(m·K) and thermal diffusivity (α) = 5×10⁻⁷ m²/s. At this condition, the midplane temperature can be obtained from heat diffusion equation for unsteady-state one-dimensional conduction:

$$\frac{T_0 - T_\infty}{T_i - T_\infty} = C_1 \exp\left(-\zeta_1^2 \frac{\alpha t}{L^2}\right)$$

where ζ_1 is the first positive root of the following equation:

$$\zeta_1 \tan \zeta_1 = hL/k$$

and C_1 can be calculated from $C_1 = \dfrac{4\sin\zeta_1}{2\zeta_1 + \sin 2\zeta_1}$

7. In a catalyst regeneration, oxygen reacts with carbon contained in an inert spherical solid matrix (catalyst pellet). Carbon is first removed from the outer edge of the solid matrix and eventually from the center core of the solid. As the carbon continues to be removed from the porous solid, the reactant gas (oxygen) must diffuse farther into the solid matrix as the reaction proceeds to reach the unreacted carbon. This process can be described using the shrinking core model.

   a. Determine the radius of carbon/oxygen interface (R) after 5000 seconds.

   b. Plot the radius of carbon/oxygen interface (R) as a function of time (t).

   For this model, the time necessary for the solid carbon interface to recede inward to a radius R is

$$t = \frac{\rho_C R_0^2 \phi}{6 D_e C_{A0}}\left[1 - 3\left(\frac{R}{R_0}\right)^2 + 2\left(\frac{R}{R_0}\right)^3\right]$$

   where ρ_C = the molar density of the carbon = 0.189 mol/cm³
   R_0 = the initial radius of carbon/oxygen interface (the outer surface of the catalyst) = 0.1 cm
   ϕ = the volume fraction of carbon in the porous catalyst = 0.5
   D_e = the effective diffusivity in the porous catalyst = 9.5 10⁻⁴ cm²/s
   C_{A0} = the concentration of oxygen at the outer surface of the porous catalyst = 1.4 10⁻⁵ mol/cm³

8. A superheated steam flows through a smooth pipe of 0.635 cm inner diameter (d_i) for a distance of 10 m (z). The temperature (T) and pressure (P_1) of the steam at the entrance are 150°C and 500 kPa, respectively. Assume that the temperature of the steam is constant. The mass flow rate of steam per unit area (G) is 23.0 kg/(s·m²).

   a. Calculate the friction factor (f) for this flow. Friction factor for turbulent flow in a smooth pipe can be obtained from:

$$f = \left[0.790 \cdot \ln(\mathrm{Re}) - 1.64\right]^{-2}$$

   where Re is the Reynolds number given by:

$$\text{Re} = \frac{d_i G}{\mu}$$

and μ is the viscosity of steam, which can be assumed constant at $13.8 \cdot 10^{-6}$ N·s/m².

b. Determine the pressure drop $(P_1 - P_2)$ of this flowing steam. For an isothermal compressible flow, the pressure drop can be obtained after solving the following equation:

$$P_1^2 - P_2^2 = f \frac{G^2 RT}{M} \frac{z}{d_i} + \frac{2G^2 RT}{M} \ln\left(\frac{P_1}{P_2}\right)$$

where P_2 is the outlet pressure, R is the gas constant and M is the molecular weight of water (= 18 kg/kmol).

c. To analyze the effect of mass flow rate of steam per unit area on the pressure drop, plot the pressure drop vs. G [$23 \leq G \leq 53$ kg/(s·m²)].

9. A sphere made of aluminum alloy with a diameter of 20 mm (D) and a surface temperature of 500°C (T_s) is suddenly immersed in a saturated water bath maintained at 100°C (T_{sat}). Since the difference between the surface temperature and the bulk temperature of the liquid is very large, a film-boiling phenomenon occurs. For this type of boiling, the heat transfer from the surface to the bulk of the liquid is governed by convection and radiation. The surface of the sphere has an emissivity of 0.25 (ε).

a. Determine the initial heat transfer coefficient due to convection (h_{conv}). A correlation that can be used to calculate this coefficient is given below:

$$\frac{h_{conv} D}{k_V} = 0.67 \left[\frac{g(\rho_L - \rho_V)\left[h_{fg} + 0.8 c p_V (T_s - T_{sat})\right] D^3}{\eta_V k_V (T_s - T_{sat})} \right]^{1/4}$$

where g is the gravitational acceleration, ρ_L is the density of liquid water [= 712.1 kg/m³], ρ_V is the density of water vapor [= 45.98 kg/m³], k_V is the thermal conductivity of water vapor [= 0.0767 W/(m·K)], cp_V is the specific heat of water vapor [= 5889 J/(kg·K)], η_V is the kinematic viscosity of water vapor [= 4.33·10⁻⁷ m²/s], and h_{fg} is the heat of vaporization of water [= 1.406·10⁶ J/kg].

b. Determine the initial total heat transfer coefficient of this film boiling (h). This coefficient can be obtained from the following equation:

$$h^{4/3} = h_{conv}^{4/3} + h_{rad} h^{1/3}$$

where h_{rad} is the initial heat transfer coefficient due to radiation, which can be calculated from

$$h_{rad} = \varepsilon\sigma\left(T_s + T_{sat}\right)\left(T_s^2 + T_{sat}^2\right)$$

In the equation above, σ is the Stefan-Boltzmann constant [= $5.67 \cdot 10^{-8}$ W/(m$^2\cdot$K^4)].

  c. To investigate the effect of the surface temperature of the sphere on the initial convective and total heat transfer coefficients of this film-boiling phenomena, plot on the same graph h_{conv} vs. T_s and h vs. T_s (400°C $\leq T_s \leq$ 600°C).

10. A tray tower is used to absorb ethyl alcohol from an inert gas stream at constant temperature and pressure using liquid water as the absorbent. The flow rate of the gas stream entering the tower (V) is 100 kmol/h and it contains 2.2 mol% alcohol ($y_2 = 0.022$). The absorbent entering the tower contains 1 mol% alcohol ($x_0 = 0.01$). The number of theoretical trays (N) of the tower is 4. The desired concentration of alcohol in the gas stream leaving the tower is 1.2 mol% ($y_1 = 0.012$).

  a. Determine the required molar flow rate of absorbent entering the tower (L).

  The required molar flow rate of absorbent can be obtained from the Kremser equation:

$$A^N = \frac{y_2 - mx_0}{y_1 - mx_0}\left(1 - \frac{1}{A}\right) + \frac{1}{A} \tag{1}$$

  where m is the thermodynamic constant (= 0.68) and A is the separation factor given by

$$A = \frac{L}{mV}$$

  Note that $A = 1$ is the trivial solution of Equation (1). Do you know why?

  b. To analyze the effect of the desired concentration of alcohol in the gas stream leaving the tower on the required molar flow rate of absorbent for $x_0 = 0.01$, plot L vs. y_1 (0.01 $\leq y_1 \leq$ 0.014).

  c. Plot L vs. x_0 (0 $\leq x_0 \leq$ 0.013) to analyze the effect of the concentration of alcohol in the absorbent entering the tower on the required molar flow rate of absorbent for $y_1 = 0.012$. Is it beneficial to use a pure absorbent?

Chapter 4
System of Equations

In chemical engineering, system of equations normally arises in many applications, including unit operations, heat transfer, reactor, material and energy balances, reaction equilibrium, and phase equilibrium calculations. If a system with n variables is described by n independent equations in these variables, then all of these n variables can in principle be calculated. If the independent equations are *all* linear (system of linear equations), there is one and only one solution. On the other hand, if any of the equations are nonlinear, there may be more than one solution or no solution at all.

4.1 System of Linear Equations

In Mathcad, a system of linear equations can be solved using two built-in functions, i.e., *lsolve* and *find*. In fact, *find* function with a *Given* block (solve block) can solve not only a system of linear equations, but also a system of non-linear equations with or without constraints, as we will see in the next two sections. Procedures 4.1 and 4.2 show how to use *lsolve* and *find* functions, respectively.

As seen in Procedure 4.1 and 4.2, *find* function is easier to use than *lsolve* function, the approach of which is based on the matrix inversion. However, *find* function cannot be used directly in the command lines of a Mathcad program (Chapter 9).

The *lsolve* function does not fully support units because it cannot be used when the unknowns have units of different types. The *find* function, on the other hand, support units, but it should be used with a certain rule, as described in the additional notes below.

Procedure 4.1: *lsolve*

To use *lsolve*, we have to set up the n equations in the matrix form:

$$\begin{bmatrix} a_{11} & a_{12} & \cdot & \cdot & a_{1n} \\ a_{21} & a_{22} & \cdot & \cdot & a_{2n} \\ \cdot & & \cdot & \cdot & \cdot \\ \cdot & & \cdot & \cdot & \cdot \\ a_{n1} & a_{n2} & \cdot & \cdot & a_{nn} \end{bmatrix} \begin{bmatrix} x_1 \\ x_2 \\ \cdot \\ \cdot \\ x_n \end{bmatrix} = \begin{bmatrix} c_1 \\ c_2 \\ \cdot \\ \cdot \\ c_n \end{bmatrix} \qquad \text{or} \qquad \mathbf{Ax = C}$$

where **A** is the $n \times n$ matrix containing coefficients of the n equations, **x** is $n \times 1$ matrix (vector) containing the n variables to be calculated, and **C** is $n \times 1$ matrix (vector) containing the constants of the n equations.

For example, we want to solve these three linear equations:

$$y1 + 2 \cdot y2 = -3 \cdot y3 + 2$$

$$y2 - y3 = 1$$

$$2 \cdot y1 + y2 + 3 \cdot y3 + 2 = 0$$

1. Create matrices A and C:

$$A := \begin{pmatrix} 1 & 2 & 3 \\ 0 & 1 & -1 \\ 2 & 1 & 3 \end{pmatrix} \qquad C := \begin{pmatrix} 2 \\ 1 \\ -2 \end{pmatrix}$$

2. Calculate the unknowns using *lsolve* and assign the result to any vector variable:

$$y := \text{lsolve}(A, C) = \begin{pmatrix} -2.5 \\ 1.5 \\ 0.5 \end{pmatrix}$$

In this case, we assign it to a vector variable y, we can assign it to other variable names.

Note that in fact *lsolve* function is equivalent to the following matrix operation:

$$y := A^{-1} \cdot C = \begin{pmatrix} -2.5 \\ 1.5 \\ 0.5 \end{pmatrix}$$

A^{-1} is the inverse of matrix A

Procedure 4.2: *find*

Find function can be used for solving a system of equations either analytically or numerically.

For example, we want to solve these three linear equations:

$$y1 + 2 \cdot y2 = -3 \cdot y3 + 2$$

$$y2 - y3 = 1$$

$$2 \cdot y1 + y2 + 3 \cdot y3 + 2 = 0$$

For <u>analytical</u> calculation:

1. Construct a *Given* block where all of the equations are defined:

Given   (*Given* is a Mathcad statement, we have to type it)

$$y1 + 2 \cdot y2 = -3 \cdot y3 + 2$$

$$y2 - y3 = 1$$

$$2 \cdot y1 + y2 + 3 \cdot y3 + 2 = 0$$

We must use the Boolean equal sign, not the assignment sign (:=) or the evaluation sign (=). This bold equal sign is typed using [Ctrl =] or obtained from the Boolean Toolbar in the Math Toolbar.

Note that the equations can be written the way we want them, we do not need, for example, to collect all of the unknowns on the left sides.

2. Calculate the unknowns using *find* function and assign the results to a vector variable y:

$$y := \text{Find}(y1, y2, y3) \rightarrow \begin{pmatrix} -\dfrac{5}{2} \\ \dfrac{3}{2} \\ \dfrac{1}{2} \end{pmatrix}$$

Note that to solve a system of equations analytically using *find* function, the function should be followed by a live symbolic equal sign (an arrow), which can be obtained from the Symbolic Keyword Toolbar in the Math Toolbar.

Find function can be typed any way we want, such as *find* or *FIND*. The same for *Given* statement.

$$y1 := y_0 = -2.5 \qquad y2 := y_1 = 1.5 \qquad y3 := y_2 = 0.5$$

The result of analytical *find* function can also be assigned to a column matrix (see below).

For <u>numerical</u> calculation:

1. Provide intial guesses: $y1 := 1$ $y2 := 1$ $y3 := 1$ Initial guesses must be provided

2. Construct a *Given* block where all of the equations are defined:

Given
$$y1 + 2 \cdot y2 = -3 \cdot y3 + 2$$
$$y2 - y3 = 1$$
$$2 \cdot y1 + y2 + 3 \cdot y3 + 2 = 0$$

For solving systems of linear equations, although numerical *find* function requires initial guesses, this function is very robust, thus it is insensitive to the choice of initial guesses.

3. Calculate the unknowns using *find* function and assign the results to a column matrix:

$$\begin{pmatrix} y1 \\ y2 \\ y3 \end{pmatrix} := Find(y1, y2, y3) = \begin{pmatrix} -2.5 \\ 1.5 \\ 0.5 \end{pmatrix}$$

The results can be assigned to either a column matrix or a vector/array variable, e.g., y.

Notes:

1. The arguments of numerical *find* function, which are the variables to be found, can be scalar (as in Procedure 4.2) and/or array variables (as in Example 2 of Example set 4.1), but not elements of an array (indexed variables), e.g., y_1, y_2, etc. On the other hand, the analytical *find* function can only take scalar variables as arguments.

2. If the unknowns have units of different types, the results must be assigned to a column matrix.

3. In a parametric solve block (parametric problem), when the unknowns have units of the same type, for numerical *find* function, the results can be assigned to either a vector function with parameters as the arguments, e.g., $y(p_1, p_2)$ or a column matrix, each element of which is a function of the parameters, e.g.,

$$\begin{pmatrix} y1(p_1, p_2) \\ y2(p_1, p_2) \end{pmatrix}$$

When the unknowns have units of different types, of course, the results can only be assigned to a column matrix. On the other hand, the analytical *find* function can only solve parametric problems when the unknowns have units of the same type and the results must be assigned to a vector function with parameters as the arguments.

Example set 4.1

1. A mixture containing 50 wt% toluene (T), 35 wt% benzene (B), and 15 wt% xylene (X) is fed to a distillation column. The top product from the column contains 6.3 wt% toluene, 91.4 wt% benzene, and 2.3 wt% xylene. The bottom product is fed to a second column. The top product from the second column contains 91.6 wt% toluene, 4.25 wt% benzene, and 4.15 wt% xylene. 10 wt% of the toluene fed to the process is recovered in the bottom product from the second column, and 83.3 wt% of the xylene fed to the process is recovered in the same stream. Determine the compositions and flow rates of all streams if the feed flow rate is 100 kg/minute.

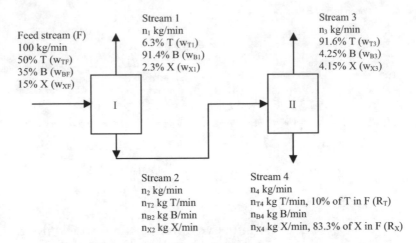

Feed stream (F)
100 kg/min
50% T (w_{TF})
35% B (w_{BF})
15% X (w_{XF})

Stream 1
n_1 kg/min
6.3% T (w_{T1})
91.4% B (w_{B1})
2.3% X (w_{X1})

Stream 3
n_3 kg/min
91.6% T (w_{T3})
4.25% B (w_{B3})
4.15% X (w_{X3})

Stream 2
n_2 kg/min
n_{T2} kg T/min
n_{B2} kg B/min
n_{X2} kg X/min

Stream 4
n_4 kg/min
n_{T4} kg T/min, 10% of T in F (R_T)
n_{B4} kg B/min
n_{X4} kg X/min, 83.3% of X in F (R_X)

From mass balance, the following system of linear equations can be set up:

$$w_{TF} \, F = w_{T1} \, n_1 + n_{T2} \qquad w_{XF} \, F = w_{X1} \, n_1 + n_{X2} \qquad n_{B2} = w_{B3} \, n_3 + n_{B4}$$
$$w_{BF} \, F = w_{B1} \, n_1 + n_{B2} \qquad n_{T2} = w_{T3} \, n_3 + n_{T4} \qquad n_{X2} = w_{X3} \, n_3 + n_{X4}$$

where $n_{T4} = R_T \, w_{TF} \, F$ and $n_{X4} = R_X \, w_{XF} \, F$

Solution: $R_T := 0.1 \qquad R_X := 0.833 \qquad F := 100$

$w_{TF} := 0.5 \qquad w_{T1} := 0.063 \qquad w_{T3} := 0.916 \qquad w_{BF} := 0.35 \qquad w_{B1} := 0.914 \qquad w_{B3} := 0.0425$

$w_{XF} := 1 - w_{TF} - w_{BF} \qquad w_{X1} := 1 - w_{T1} - w_{B1} \qquad w_{X3} := 1 - w_{T3} - w_{B3}$

$n_{T4} := R_T \cdot w_{TF} \cdot F \qquad n_{X4} := R_X \cdot w_{XF} \cdot F$

Given

$$w_{TF} \cdot F = w_{T1} \cdot n_1 + n_{T2} \qquad w_{XF} \cdot F = w_{X1} \cdot n_1 + n_{X2} \qquad n_{B2} = w_{B3} \cdot n_3 + n_{B4}$$
$$w_{BF} \cdot F = w_{B1} \cdot n_1 + n_{B2} \qquad n_{T2} = w_{T3} \cdot n_3 + n_{T4} \qquad n_{X2} = w_{X3} \cdot n_3 + n_{X4}$$

$$N := \text{Find}\left(n_1, n_{T2}, n_{B2}, n_{X2}, n_3, n_{B4}\right) \text{ float}, 5 \; \rightarrow \begin{pmatrix} 23.144 \\ 48.542 \\ 13.847 \\ 14.468 \\ 47.535 \\ 11.827 \end{pmatrix}$$

$n_1 := N_0 \qquad n_{T2} := N_1 \qquad n_{B2} := N_2 \qquad n_{X2} := N_3 \qquad n_3 := N_4 \qquad n_{B4} := N_5$

The flow rate of stream 1 (kg/min): $n_1 = 23.144$

The flow rate of stream 2 (kg/min): $n_2 := n_{T2} + n_{B2} + n_{X2} = 76.857$

Composition of stream 2:

$$w_{T2} := \frac{n_{T2}}{n_2} = 0.632 \qquad w_{B2} := \frac{n_{B2}}{n_2} = 0.18$$

$$w_{X2} := 1 - w_{T2} - w_{B2} = 0.188$$

The flow rate of stream 3 (kg/min): $n_3 = 47.535$

The flow rate of stream 4 (kg/min): $n_4 := n_{T4} + n_{B4} + n_{X4} = 29.322$

Composition of stream 4:

$$w_{T4} := \frac{n_{T4}}{n_4} = 0.171 \qquad w_{B4} := \frac{n_{B4}}{n_4} = 0.403$$

$$w_{X4} := 1 - w_{T4} - w_{B4} = 0.426$$

Solution using *Isolve* function:

$$M := \begin{pmatrix} w_{T1} & 1 & 0 & 0 & 0 & 0 \\ w_{B1} & 0 & 1 & 0 & 0 & 0 \\ w_{X1} & 0 & 0 & 1 & 0 & 0 \\ 0 & -1 & 0 & 0 & w_{T3} & 0 \\ 0 & 0 & -1 & 0 & w_{B3} & 1 \\ 0 & 0 & 0 & -1 & w_{X3} & 0 \end{pmatrix} \qquad C := \begin{pmatrix} w_{TF} \cdot F \\ w_{BF} \cdot F \\ w_{XF} \cdot F \\ -n_{T4} \\ 0 \\ -n_{X4} \end{pmatrix}$$

$$\begin{pmatrix} n_1 \\ n_{T2} \\ n_{B2} \\ n_{X2} \\ n_3 \\ n_{B4} \end{pmatrix} := \text{Isolve}(M,C) = \begin{pmatrix} 23.144 \\ 48.542 \\ 13.847 \\ 14.468 \\ 47.535 \\ 11.827 \end{pmatrix}$$

2. A long column, 1 m by 1 m on a side, is exposed to a hot air such that the temperatures of three surfaces (T_s) of the column are maintained at 600 K while the remaining surface is exposed to a colder air at 300 K (T_a) with a convective heat transfer coefficient (h) of 16 W/(m²K). Estimate the two-dimensional temperature distribution in the column at steady state if the conductivity (k) of the column material is 1 W/(m K).

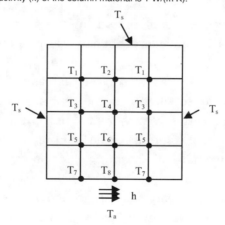

Using the energy balance method, we can write the following finite-difference equations with a grid size (Δx) of 0.25 m:

$$4 \cdot T1 = T2 + T3 + 2 \cdot T_s \qquad 4 \cdot T3 = T1 + T4 + T5 + T_s \qquad 4 \cdot T5 = T3 + T6 + T7 + T_s$$

$$4 \cdot T2 = 2 \cdot T1 + T4 + T_s \qquad 4 \cdot T4 = T2 + 2 \cdot T3 + T6 \qquad 4 \cdot T6 = T4 + 2 \cdot T5 + T8$$

$$2 \cdot \left(\frac{h \cdot \Delta x}{k} + 2 \right) \cdot T7 = 2 \cdot T5 + T8 + T_s + 2 \cdot \frac{h \cdot \Delta x}{k} \cdot T_a$$

$$2 \cdot \left(\frac{h \cdot \Delta x}{k} + 2 \right) \cdot T8 = 2 \cdot T6 + 2 \cdot T7 + 2 \cdot \frac{h \cdot \Delta x}{k} \cdot T_a$$

Solution: ORIGIN:= 1

$$\Delta x := 0.25 \, m \qquad h := 16 \cdot \frac{W}{m^2 K} \qquad k := 1 \cdot \frac{W}{m \cdot K} \qquad T_s := 600 \, K \qquad T_a := 300 \, K$$

Initial guess: $\quad i := 1..8 \qquad T_i := 500 \cdot K$

Given

$$4 \cdot T_1 = T_2 + T_3 + 2 \cdot T_s \qquad 4 \cdot T_4 = T_2 + 2 \cdot T_3 + T_6$$

$$4 \cdot T_2 = 2 \cdot T_1 + T_4 + T_s \qquad 4 \cdot T_5 = T_3 + T_6 + T_7 + T_s$$

$$4 \cdot T_3 = T_1 + T_4 + T_5 + T_s \qquad 4 \cdot T_6 = T_4 + 2 \cdot T_5 + T_8$$

$$2 \cdot \left(\frac{h \cdot \Delta x}{k} + 2 \right) \cdot T_7 = 2 \cdot T_5 + T_8 + T_s + 2 \cdot \frac{h \cdot \Delta x}{k} \cdot T_a$$

$$2 \cdot \left(\frac{h \cdot \Delta x}{k} + 2 \right) \cdot T_8 = 2 \cdot T_6 + 2 \cdot T_7 + 2 \cdot \frac{h \cdot \Delta x}{k} \cdot T_a$$

$$T := \text{Find}(T) = \begin{pmatrix} 582.294 \\ 575.476 \\ 553.701 \\ 537.316 \\ 495.194 \\ 466.387 \\ 360.686 \\ 337.846 \end{pmatrix} K$$

4.2 System of Non-Linear Equations

In Mathcad, a system of non-linear equations can be solved using *find* function. The procedure how to use this function has been described in the previous section. As in solving systems of linear equations, *find* function can be used to solve either analytically or numerically. However, unlike in solving systems of linear equations, the use of analytical *find* function to solve a system of non-linear equations can give us more than one set of solutions or no solution at all. Whether or not all solutions are physically important depends on the system in question. If analytical *find* function does not give us any solution, especially if the number of equations is large and the equations are very complex, try to use *find* function to solve numerically. Remember that a system of non-linear equations may not have any solution. If the unknowns have units of different types, numerical *find* function must be used and the results must be assigned to a column matrix.

Numerical *find* function is also recommended for a parametric problem (parametric solve block). In this case, as also described in the previous section, when the unknowns have units of the same type, the results can be assigned to either a vector function with parameters as the arguments or a column matrix, each element of which is a function of the parameters. When the unknowns have units of different types, of course, the results can only be assigned to a column matrix.

Unlike in solving a system of linear equations, the success of numerical *find* function in solving a system of non-linear equations is highly dependent on the choice of initial guesses. Reasonable initial guesses for a certain problem could usually be obtained from the physical consideration of the problem. For example, if the variables represent mole fractions, we know that the values of those variables are from 0 to 1. If *find* function does not converge, one may try to change the numerical method used by right clicking on the *find* function and choosing the desired method or to change the initial guesses.

Example set 4.2

1. Ethylene and acetylene are produced by dehydrogenation of ethane at 977°C and 1 atm in a reactor, where the following catalytic reactions take place:

$$\text{Reaction 1: } C_2H_6 \Leftrightarrow C_2H_4 + H_2$$

$$\text{Reaction 2: } C_2H_6 \Leftrightarrow C_2H_2 + 2H_2$$

The equilibrium constant of reaction 1 (K_1) is 3.75 and the equilibrium constant of reaction 2 (K_2) is 0.135. If the residence time of the reacting mixture in the reactor is long enough that the equilibrium is achieved, determine the composition of the product stream.

The equilibrium constants K_1 and K_2 can be expressed in terms of the extents of the two reactions (ξ_1 and ξ_2):

$$K_1 = \frac{\xi_1 \cdot (\xi_1 + 2 \cdot \xi_2)}{(100 - \xi_1 - \xi_2) \cdot (100 + \xi_1 + 2 \cdot \xi_2)} \qquad K_2 = \frac{\xi_2 \cdot (\xi_1 + 2 \cdot \xi_2)^2}{(100 - \xi_1 - \xi_2) \cdot (100 + \xi_1 + 2 \cdot \xi_2)^2}$$

and the mol fraction of each species in the product stream can be obtained from:

$$x_{C2H6} = \frac{100 - \xi_1 - \xi_2}{100 + \xi_1 + 2 \cdot \xi_2} \qquad x_{C2H4} = \frac{\xi_1}{100 + \xi_1 + 2 \cdot \xi_2}$$

$$x_{C2H2} = \frac{\xi_2}{100 + \xi_1 + 2 \cdot \xi_2} \qquad x_{H2} = \frac{\xi_1 + 2 \cdot \xi_2}{100 + \xi_1 + 2 \cdot \xi_2}$$

Solution: ORIGIN≡ 1

Given

$$\frac{\xi_1 \cdot (\xi_1 + 2 \cdot \xi_2)}{(100 - \xi_1 - \xi_2) \cdot (100 + \xi_1 + 2 \cdot \xi_2)} = 3.75 \qquad \frac{\xi_2 \cdot (\xi_1 + 2 \cdot \xi_2)^2}{(100 - \xi_1 - \xi_2) \cdot (100 + \xi_1 + 2 \cdot \xi_2)^2} = 0.135$$

$$x := \text{Find}(\xi_1, \xi_2) \text{ float}, 5 \rightarrow \begin{pmatrix} 200.0 & 214.53 & -89.602 & 83.063 \\ -100.0 & -110.53 & 0.40738 & 6.1274 \end{pmatrix} \quad \text{There are four sets of solutions}$$

Since the extent of reactions are positive numbers, we choose only the fourth set in the last column:

$$\begin{pmatrix} \xi_1 \\ \xi_2 \end{pmatrix} := x^{\langle 4 \rangle} \qquad \text{The Matrix Column operator (M}^{\langle\rangle}\text{) for extracting a column from a matrix can be accessed from the Matrix Toolbar}$$

Composition of the product stream:
$$x_{C2H6} := \frac{100 - \xi_1 - \xi_2}{100 + \xi_1 + 2 \cdot \xi_2} \qquad x_{C2H4} := \frac{\xi_1}{100 + \xi_1 + 2 \cdot \xi_2}$$

$$x_{C2H2} := \frac{\xi_2}{100 + \xi_1 + 2 \cdot \xi_2} \qquad x_{H2} := \frac{\xi_1 + 2 \cdot \xi_2}{100 + \xi_1 + 2 \cdot \xi_2}$$

$$x_{C2H6} = 0.055 \qquad x_{C2H4} = 0.425 \qquad x_{C2H2} = 0.031 \qquad x_{H2} = 0.488$$

2. A liquid mixture containing 50 mol% n-pentane (C5) and 50 mol% n-heptane (C7) at a high pressure enters a flash vaporizer at a lower presure. The pressure and temperature of this flash vaporizer can be controlled as needed. The two product streams, i.e., a vapor stream and a liquid stream, coming out from the vaporizer are in equilibrium. At the condition of interest, both liquid and vapor can be assumed ideal, thus Raoult's law can be applied. The vapor pressures of C5 and C7 at temperature t (°C) can be obtained using Antoine equation:

$$\ln\left(Psat_i\right) \;=\; A_i - \frac{B_i}{t + C_i}$$

$Psat_i$ = vapor pressure of component i [kPa]
A_i, B_i, C_i = constants for component i

Data: A_{C5} = 13.8183 B_{C5} = 2477.07 C_{C5} = 233.21
            A_{C7} = 13.8587 B_{C7} = 2991.32 C_{C7} = 216.64

a. Determine the maximum operating pressure (P) if the temperature (T) of the vaporizer is set to 330 K and at least 70% of the feed should be vaporized (v). Also, calculate the composition in each product stream (x_{C5}, x_{C7}, y_{C5}, y_{C7}) at the operating pressure.
b. Plot the operating pressure as a function of the fraction of vapor vaporized at 330K (0<v<1).

The following equations can be set up:

C5 mol balance $z_{C5} = y_{C5}\, v + x_{C5}\,(1-v)$

Raoult's law $y_{C5}\, P = x_{C5}\, Psat_{C5}$

            $(1-y_{C5})\, P = (1-x_{C5})\, Psat_{C7}$

where z_{C5} is the mol fraction of C5 in the feed stream, x_{C5} is the mol fraction of C5 in the liquid stream, y_{C5} is the mol fraction of C5 in the vapor stream, and v is the molar fraction of feed that is vaporized.

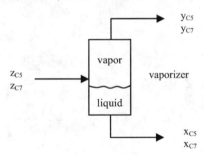

Solution: Coefficients of Antoine equation: $A_{C5} := 13.8183$ $B_{C5} := 2477.07$ $C_{C5} := 233.21$
                                    $A_{C7} := 13.8587$ $B_{C7} := 2991.32$ $C_{C7} := 216.64$

$t := 330 - 273$

$$Psat_{C5} := \exp\left(A_{C5} - \frac{B_{C5}}{t + C_{C5}} \right)\cdot kPa \qquad Psat_{C7} := \exp\left(A_{C7} - \frac{B_{C7}}{t + C_{C7}} \right)\cdot kPa$$

The mol fraction of C5 in the feed: $z_{C5} := 0.5$

Initial guesses: $y_{C5} := 0.8$ $x_{C5} := 0.1$ $P := 80\cdot kPa$

Given

            $z_{C5} = y_{C5}\cdot v + x_{C5}\cdot(1-v)$

            $y_{C5}\cdot P = x_{C5}\cdot Psat_{C5}$

            $\left(1 - y_{C5}\right)\cdot P = \left(1 - x_{C5}\right)\cdot Psat_{C7}$

$$\begin{pmatrix} y_{C5}(v) \\ x_{C5}(v) \\ P(v) \end{pmatrix} := Find\left(y_{C5}, x_{C5}, P\right)$$

a. The maximum pressure when at least 70% of the feed should be vaporized:

$P(0.7) = 45.374 \text{kPa}$

b. Plot P vs. v: $v := 0, 0.1 .. 1$

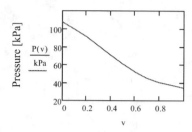

3. a. Determine the bubble point pressure (P) of a liquid mixture containing 25 mol% chloroform (x_1) and 75 mol% ethanol at 55°C. Also, calculate the mol fraction of chloroform in the vapor phase (y_1) at that bubble point pressure. From thermodynamics, the following equations can be set up:

$$y_1 \hat{\phi}_1 P = x_1 \gamma_1 Psat_1$$

$$(1 - y_1) \hat{\phi}_2 P = (1 - x_1) \gamma_2 Psat_2$$

where y_1 is the mol fraction of component 1 in the vapor phase, x_1 is the mol fraction of component 1 in the liquid phase, $Psat_i$ is the vapor pressure of component i, γ_i is the activity coefficient of component i, and $\hat{\phi}_i$ is the fugacity coefficient of component i calculated from:

$$RT \ln \hat{\phi}_1 = B_{11}(P - Psat_1) + P(1 - y_1)^2 \delta_{12}$$

$$RT \ln \hat{\phi}_2 = B_{22}(P - Psat_2) + P y_1^2 \delta_{12}$$

$$\delta_{12} = 2B_{12} - B_{11} - B_{22}$$

where R is the gas constant and T is the temperature in K. The activity coefficients are calculated from a Gamma model:

$$\ln \gamma_1 = (1 - x_1)^2 [A_{12} + 2(A_{21} - A_{12})x_1] \qquad \ln \gamma_2 = x_1^2 [A_{21} + 2(A_{12} - A_{21})(1 - x_1)]$$

Data: $Psat_1 = 82.37$ kPa, $Psat_2 = 37.31$ kPa

$B_{11} = -963$ cm³/mol, $B_{22} = -1523$ cm³/mol, $B_{12} = 52$ cm³/mol
$A_{12} = 0.59$, $A_{21} = 1.42$

b. Plot the bubble point pressure as a function of the mol fraction of component 1 in the liquid phase. On the same graph, also plot the bubble point pressure as a function of the mol fraction of component 1 in the vapor phase.

Solution: $R := 8.314472 \dfrac{\text{joule}}{\text{mole} \cdot \text{K}}$ $T := 55°C$

$A_{12} := 0.59$ $A_{21} := 1.42$ $Psat_1 := 82.37 \text{kPa}$ $Psat_2 := 37.31 \cdot \text{kPa}$

$$\gamma_1(x_1) := \exp\left[(1 - x_1)^2 \cdot \left[A_{12} + 2 \cdot (A_{21} - A_{12}) \cdot x_1\right]\right]$$

$$\gamma_2(x_1) := \exp\left[x_1^2 \cdot \left[A_{21} + 2 \cdot (A_{12} - A_{21}) \cdot (1 - x_1)\right]\right]$$

$B_{11} := -963 \cdot \dfrac{\text{cm}^3}{\text{mol}}$ $B_{22} := -1523 \cdot \dfrac{\text{cm}^3}{\text{mol}}$ $B_{12} := 52 \cdot \dfrac{\text{cm}^3}{\text{mol}}$ $\delta_{12} := 2 \cdot B_{12} - B_{11} - B_{22}$

$$\Phi_1(P, y_1) := \exp\left[\dfrac{B_{11} \cdot (P - Psat_1) + P \cdot (1 - y_1)^2 \cdot \delta_{12}}{R \cdot T}\right]$$

$$\Phi_2(P, y_1) := \exp\left[\dfrac{B_{22} \cdot (P - Psat_2) + P \cdot y_1^2 \cdot \delta_{12}}{R \cdot T}\right]$$

Initial guesses: $y_1 := 0.2$ $P := \dfrac{Psat_1 + Psat_2}{2}$

Given

$$y_1 \cdot \Phi_1(P, y_1) \cdot P = x_1 \cdot \gamma_1(x_1) \cdot Psat_1$$
$$(1 - y_1) \cdot \Phi_2(P, y_1) \cdot P = (1 - x_1) \cdot \gamma_2(x_1) \cdot Psat_2$$

$$\begin{pmatrix} y_1(x_1) \\ P(x_1) \end{pmatrix} := Find(y_1, P)$$

a. The bubble pressure for $x_1 = 0.25$: $P(0.25) = 63.757 kPa$

The mole fraction of chloroform in the vapor phase for $x_1 = 0.25$: $y_1(0.25) = 0.558$

b. $x_1 := 0, 0.01 .. 1$

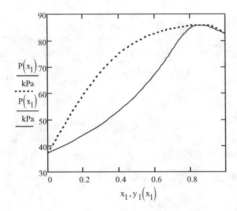

4.3 System of Equations with Constraints

As mentioned in the previous section, a system of non-linear equations could give us more than one set of solutions. In Example 1 of Example set 4.2, we can solve the system of equations analytically and easily know which the correct set is because the extent of reaction cannot be a negative number. For problems where we cannot solve analytically, selecting the correct solution is not always trivial. Therefore, it is better to include constraints, if any, in the *Given* block to exclude any possible solutions that are not physical. The constraints are usually dictated by the physical problem of interest.

A system of equations with constraints in a *Given* block can only be solved numerically with *find* function. *Find* function for analytical calculations does not allow any inequalities in the *Given* block.

Example set 4.3

1. Two simultaneous reactions occur at 900 K and 1 atm:

Reaction A: $2CO + 2H_2 \Leftrightarrow CO_2 + CH_4$

Reaction B: $CO + 3H_2 \Leftrightarrow H_2O + CH_4$

At this condition, the equilibrium constants of these two reactions are:

$$K_A = 1.8283 \qquad K_B = 0.7902$$

Determine the equilibrium composition of the gas mixture if the initial numbers of moles of CO (n_1) and H_2 (n_2) are 2 and 1, respectively.

The following two equations can be set up from the definition of equilibrium constant:

$$K_A = \frac{\xi_A \cdot (\xi_A + \xi_B) \cdot (n_1 + n_2 - 2 \cdot \xi_A - 2 \cdot \xi_B)^2}{(n_1 - 2 \cdot \xi_A - \xi_B)^2 \cdot (n_2 - 2 \cdot \xi_A - 3 \cdot \xi_B)^2} \qquad K_B = \frac{\xi_B \cdot (\xi_A + \xi_B) \cdot (n_1 + n_2 - 2 \cdot \xi_A - 2 \cdot \xi_B)^2}{(n_1 - 2 \cdot \xi_A - \xi_B) \cdot (n_2 - 2 \cdot \xi_A - 3 \cdot \xi_B)^3}$$

where ξ_A and ξ_B are the extents of reaction A and B, respectively.

The extent of reaction should be a positive number:

$$\xi_A \geq 0 \qquad\qquad \xi_B \geq 0$$

Since the amount of reacted CO and H_2 cannot exceed their initial amount, there are two additional inequalities that can be obtained:

$$2 \cdot \xi_A + \xi_B \leq 2 \qquad 2 \cdot \xi_A + 3 \cdot \xi_B \leq 1$$

The equilibrium composition can be calculated using the following equations:

$$y_{CO} = \frac{n_1 - 2 \cdot \xi_A - \xi_B}{n_1 + n_2 - 2 \cdot \xi_A - 2 \cdot \xi_B} \qquad y_{H2} = \frac{n_2 - 2 \cdot \xi_A - 3 \cdot \xi_B}{n_1 + n_2 - 2 \cdot \xi_A - 2 \cdot \xi_B}$$

$$y_{CO2} = \frac{\xi_A}{n_1 + n_2 - 2 \cdot \xi_A - 2 \cdot \xi_B} \qquad y_{CH4} = \frac{\xi_A + \xi_B}{n_1 + n_2 - 2 \cdot \xi_A - 2 \cdot \xi_B}$$

$$y_{H2O} = \frac{\xi_B}{n_1 + n_2 - 2 \cdot \xi_A - 2 \cdot \xi_B}$$

Solution:

Let us first solve the system of equations without imposing any constraints:

$$n_1 := 2 \qquad n_2 := 1$$

Initial guesses: $\xi_A := 0.05 \qquad \xi_B := 0.5$

Given

$$\frac{\xi_A \cdot (\xi_A + \xi_B) \cdot (n_1 + n_2 - 2 \cdot \xi_A - 2 \cdot \xi_B)^2}{(n_1 - 2 \cdot \xi_A - \xi_B)^2 \cdot (n_2 - 2 \cdot \xi_A - 3 \cdot \xi_B)^2} = 1.8283$$

$$\frac{\xi_B \cdot (\xi_A + \xi_B) \cdot (n_1 + n_2 - 2 \cdot \xi_A - 2 \cdot \xi_B)^2}{(n_1 - 2 \cdot \xi_A - \xi_B) \cdot (n_2 - 2 \cdot \xi_A - 3 \cdot \xi_B)^3} = 0.7902$$

$$\begin{pmatrix} \xi_A \\ \xi_B \end{pmatrix} := \text{Find}(\xi_A, \xi_B) = \begin{pmatrix} 0.038 \\ 1.966 \end{pmatrix}$$

This solution sounds reasonable because the extents of reaction are positive numbers.

Now, let us include the constraints in the *Given* block and use the same initial guesses:

Initial guesses: $\xi_A := 0.05 \qquad \xi_B := 0.5$

Given

$$\frac{\xi_A \cdot (\xi_A + \xi_B) \cdot (n_1 + n_2 - 2 \cdot \xi_A - 2 \cdot \xi_B)^2}{(n_1 - 2 \cdot \xi_A - \xi_B)^2 \cdot (n_2 - 2 \cdot \xi_A - 3 \cdot \xi_B)^2} = 1.8283$$

$$\frac{\xi_B \cdot (\xi_A + \xi_B) \cdot (n_1 + n_2 - 2 \cdot \xi_A - 2 \cdot \xi_B)^2}{(n_1 - 2 \cdot \xi_A - \xi_B) \cdot (n_2 - 2 \cdot \xi_A - 3 \cdot \xi_B)^3} = 0.7902$$

$$\xi_A \geq 0$$
$$\xi_B \geq 0$$
$$2 \cdot \xi_A + \xi_B \leq 2$$
$$2 \cdot \xi_A + 3 \cdot \xi_B \leq 1$$

The inequality signs can be accessed from the Boolean Toolbar.

$$\begin{pmatrix} \xi_A \\ \xi_B \end{pmatrix} := \text{Find}(\xi_A, \xi_B) = \begin{pmatrix} 0.275 \\ 0.03 \end{pmatrix}$$

This is the correct solution.

$$y_{CO} := \frac{n_1 - 2 \cdot \xi_A - \xi_B}{n_1 + n_2 - 2 \cdot \xi_A - 2 \cdot \xi_B}$$

$$y_{H2} := \frac{n_2 - 2 \cdot \xi_A - 3 \cdot \xi_B}{n_1 + n_2 - 2 \cdot \xi_A - 2 \cdot \xi_B}$$

$$y_{CO2} := \frac{\xi_A}{n_1 + n_2 - 2 \cdot \xi_A - 2 \cdot \xi_B}$$

$$y_{CH4} := \frac{\xi_A + \xi_B}{n_1 + n_2 - 2 \cdot \xi_A - 2 \cdot \xi_B}$$

$$y_{H2O} := \frac{\xi_B}{n_1 + n_2 - 2 \cdot \xi_A - 2 \cdot \xi_B}$$

Thus, the composition of the gas mixture is:

$y_{CO} = 0.594$ $y_{H2} = 0.151$ $y_{CO2} = 0.115$

$y_{CH4} = 0.128$ $y_{H2O} = 0.013$

Problems

1. a. Determine the bubble point temperature (t) of a liquid binary mix-
        ture containing 20 mol% acetone (x_1) and 80 mol% methanol at 1
        atm (P). Also, calculate the mol fraction of acetone in the vapor
        phase (y_1) at that bubble point temperature. From thermodynam-
        ics, the following equations can be set up:

$$y_1 P = x_1 \gamma_1 Psat_1$$
$$(1 - y_1)P = (1 - x_1)\gamma_2 Psat_2$$

where $Psat_i$ is the vapor pressure of component i and γ_i is the activity coefficient of component i.

The vapor pressure ($Psat$ in kPa) at $t°C$ can be calculated from Antoine equation:

$$\ln Psat_i = A_i - \frac{B_i}{t + C_i}$$

where t is the temperature and A_i, B_i, and C_i are the constants for
component i given below:

	A	B	C
Acetone (1)	14.3916	2795.82	230.00
Methanol (2)	16.5938	3644.30	239.76

The activity coefficient can be calculated from a Gamma model:

$$\gamma_1 = \exp\left(0.64(1 - x_1)^2\right) \qquad \gamma_2 = \exp\left(0.64 x_1^2\right)$$

   b. Plot the bubble point temperature as a function of the mol fraction
       of component 1 in the liquid phase (x_1). Note: y-axis: t, x-axis: x_1.
   c. Create a new graph to plot the bubble point temperature, which is
       a function of x_1, vs. the mol fraction of component 1 in the vapor
       phase (y_1), which is also a function of x_1. Note: y-axis: t, x-axis: y_1.

Then, include t vs. x_l as in part b on this graph so that you have two curves.

2. A saturated solution of AgNO₃ at 100°C is cooled to 20°C in a crystallizer and the crystals formed are filtered out. The wet filter cake contains 75 weight% solid crystals (w_c) and 25 weight% saturated solution. The remaining water in the wet cake is then removed in a dryer.

   a. Determine the mass of AgNO₃ eventually recovered (R) per 100 kg saturated solution fed to the crystallizer (F).

   b. Plot the mass of AgNO₃ eventually recovered (R) per 100 kg saturated solution fed to the crystallizer as a function of the percentage of solid crystals in the wet filter cake (w_c). Why does the mass of AgNO₃ recovered decrease with increasing percentage of solid crystals in the filter cake? Can the wet filter cake contain 65 weight% solid crystals? Explain.

Solubility data:

Temperature (°C)	Solubility of AgNO₃ (g/g solution)
20	0.689 (s_{20})
100	0.905 (s_{100})

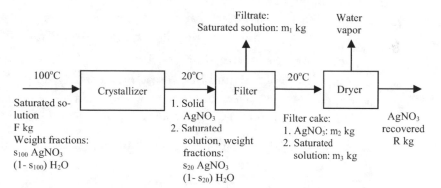

We obtain the following equations from material balance:

$$m_2 = \frac{w_c}{1 - w_c} m_3$$

$$(1 - s_{100})F = (1 - s_{20})m_1 + (1 - s_{20})m_3$$

$$F = m_1 + m_2 + m_3$$

The mass of AgNO₃ eventually recovered is then calculated from:

$$R = s_{100}F - s_{20}m_1$$

3. A first order irreversible liquid reaction will be carried out in a continuous-stirred tank reactor (CSTR). The reaction is as follows:

$$A \rightarrow products \qquad -r_A = kC_A \qquad k = k_0 \exp\left(-\frac{1400K}{T}\right)$$

where $-r_A$ is the reaction rate [mol/(cm³.hr)], C_A is the concentration of A in the reactor [mol/cm³], which is the same as the concentration of A in the exit stream, k is the reaction constant, k_0 is the frequency factor (= 450 hr⁻¹), and T is the absolute temperature of the fluid in the reactor in Kelvin. The reaction is exothermic and it is suggested to use a cooling medium with a temperature of 273.15 K (T_c). A feed stream containing 0.5 mol A/cm³ (C_{A0}) will be used and the residence time of the fluid in the reactor (τ) will be 0.2 hr. General equations derived from mole and energy balances are as follows

$$\tau\frac{dC_A}{dt} = \left(C_{A0} - C_A\right) - \left(-r_A\tau\right)$$

$$\tau\frac{dT}{dt} = \left(\frac{-\Delta H_R}{c_{ps}}\right)\left(\frac{-r_A}{C_{A0}}\right)\tau - \left(1 + \kappa\right)\left(T - T_c\right)$$

where t = time (hr)

ΔH_R = heat of reaction = $-151000 + 2\cdot(T - 298.15)$ J/mol

c_{ps} = the average heat capacity of the solution = 30 J/(mol·K)

κ = the dimensionless heat transfer parameter = 80

Determine the steady-state concentration of A and the steady-state temperature of the fluid in the reactor. At steady state, the concentration of A and the temperature of the fluid in the reactor do not change with time. Thus, at steady state, $dC_A/dt = 0$ and $dT/dt = 0$.

4. A total of 20,000 scfm of air (L_f) will be treated in a membrane separator. Assume the composition of air is 79 mol% N_2 and 21 mol% O_2 (x_f). A low-density polyethylene membrane of 0.2 µm thick (t) is being considered. Assume a pressure of 1.034×10⁶ Pa on the feed (reject) side (p_h) and a pressure of 1.034×10⁵ Pa on the permeate side (p_l). If the desired cut θ (fraction of feed permeated) is 0.2, determine the area of the membrane needed (A_m) and the oxygen mol fraction in the reject stream (x_o) if complete mixing model can be used. The permeabilities of oxygen (P_A) and nitrogen (P_B) in the membrane are 2.2× 10^{-13} and 0.73×10^{-13} cm³ (STP)/(cm²·s·Pa/cm), respectively.

For complete mixing model, the following equations can be derived from material balance and mass transfer:

$$x_f = (1 - \theta)x_o + \theta y_p$$

$$\theta L_f y_p = \frac{P'_A A_m (p_h x_o - p_l y_p)}{t}$$

$$\theta L_f (1 - y_p) = \frac{P'_B A_m (p_h(1 - x_o) - p_l(1 - y_p))}{t}$$

where y_p is the oxygen mol fraction in the permeate stream.

5. A fresh feed ($F = 100$ mol/hr) containing 26 mol% N_2 (x_{N2F}), 73 mol% H_2, and 1 mol% inert (x_{IF}) is fed to an ammonia production process. Before it enters a reactor to produce ammonia, it is combined with a recycle stream (R mol/hr), which contains the same substances. This recycle stream has 10 mol% inert (x_{IR}). In the reactor, the following reaction takes place:

$$N_2 + 3H_2 \rightarrow 2NH_3$$

The conversion of N_2 (X) in the reactor is 25%. The products coming out from the reactor (S mol/hr) pass through a condenser where all the ammonia produced is condensed and separated from the remaining gases. The remaining gases are recycled after being partly purged to prevent the accumulation of the inert in the system. Determine the rate of the stream entering the reactor (M mol/hr), the recycle stream, and the ammonia produced ($x_{NH3S}S$).

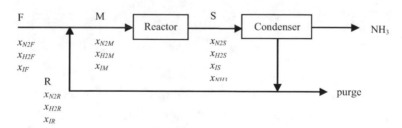

The following equations that need to be solved are obtained from material balance:

Mixing point:

$$x_{N2F}F + x_{N2R}R = x_{N2M}M$$
$$F + R = M$$
$$x_{IF}F + x_{IR}R = x_{IM}M$$

Reactor:

$$x_{N2S}S = (1 - X)x_{N2M}M$$
$$x_{H2S}S = (1 - x_{N2M} - x_{IM})M - 3Xx_{N2M}M$$

Total element balance:

$$2x_{N2F}F = x_{NH3S}S + 2x_{N2R}\frac{x_{IF}F}{x_{IR}}$$

$$2(1 - x_{N2F} - x_{IF})F = 3x_{NH3S}S + 2(1 - x_{N2R} - x_{IR})\frac{x_{IF}F}{x_{IR}}$$

Two additional equations from the definition of mol fraction:

$$x_{IR}(1 - x_{N2M}) = x_{IM}$$

$$x_{H2S} + x_{N2S} + x_{NH3S} + \frac{x_{IM}M}{S} = 1$$

6. A counterflow double pipe heat exchanger is designed to cool a process fluid entering at 100°C (t_{hi}). The flow rates of the process fluid through the annulus ($\dot{m}_h$) is 0.1 kg/s and the cooling medium ($\dot{m}_c$), which is water, through the inner tube (D_i = 25 mm) is 0.2 kg/s. The water enters at a temperature of 30°C (t_{ci}) and the length of the double pipe (L) is 3.762 m. If the overall heat transfer coefficient U_i is known to be 770 W/(m²·K), determine the outlet temperatures of the process fluid (t_{ho}) and water (t_{co}). Specific heats for the process fluid (c_h) = 2131 J/(kg·K) and for water (c_c) = 4178 J/(kg·K) From heat balance and heat transfer the following equations can be obtained:

$$U_i \pi D_i L \frac{(t_{hi} - t_{co}) - (t_{ho} - t_{ci})}{\ln\left(\dfrac{t_{hi} - t_{co}}{t_{ho} - t_{ci}}\right)} = \dot{m}_c c_c (t_{co} - t_{ci})$$

$$\dot{m}_h c_h (t_{hi} - t_{ho}) = \dot{m}_c c_c (t_{co} - t_{ci})$$

7. The minimum reflux ratio of a debutanizer shown below needs to be determined. The mole fraction of each component in the feed ($z_{F,i}$), the specified molar flow rates of components in the distillate (d_i), and the relative volatility ($\alpha_{i,3}$) are as follows:

Component	i	$z_{F,i}$	d_i [lbmol/hr]	$\alpha_{i,3}$
i-C4	1	0.0137	12	2.43
n-C4	2	0.5113	442	1.93
i-C5	3	0.0411	13	1.00
n-C5	4	0.0171	d_4	0.765
n-C6	5	0.0262	0	0.362
n-C7	6	0.0446	0	0.164
n-C8	7	0.3106	0	0.0720
n-C9	8	0.0354	0	0.0362

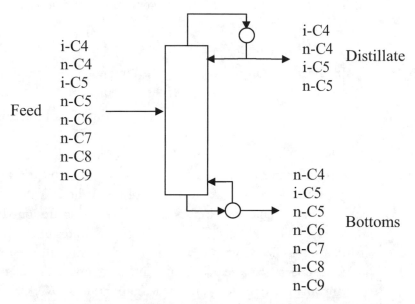

The minimum reflux ratio ($R_{min} = L_{min}/D$) can be obtained by solving Underwood equations:

$$\sum_i \frac{\alpha_{i,3} z_{F,i}}{\alpha_{i,3} - \theta_1} = 1 - q \qquad \sum_i \frac{\alpha_{i,3} z_{F,i}}{\alpha_{i,3} - \theta_2} = 1 - q$$

$$\sum_i \frac{\alpha_{i,3} d_i}{\alpha_{i,3} - \theta_1} = D + L_{min} \qquad \sum_i \frac{\alpha_{i,3} d_i}{\alpha_{i,3} - \theta_2} = D + L_{min}$$

$$\sum_i d_i = D$$

where L_{min} is the molar flow rate of liquid returned to the column at minimum reflux, D is the molar flow rate of distillate withdrawn from the column, and q is the feed quality (= 0.8666), i.e., mole fraction of liquid in the flashed feed. The two quantities θ_1 and θ_2 must satisfy the following constraint:

$$\alpha_{2,3} > \theta_1 > \alpha_{3,3} > \theta_2 > \alpha_{4,3}$$

Solve the above system of non-linear equations and calculate the minimum reflux ratio.

8. An exothermic reaction is carried out in an adiabatic continuous-stirred tank reactor with a fluid residence time of 0.27 hours (τ). The standard heat of reaction (ΔH_R^0 at T_R = 293.15 K) for this reaction is $-1.216 \cdot 10^5$ J/mol.

a. From material balance, the conversion (X) of reactant is given by:

$$X = \frac{k\tau}{1 + k\tau} \qquad (1)$$

In the equation above, k is the temperature-dependent reaction constant correlated by the *Arrhenius* equation:

$$k = Ae^{-\frac{E}{RT}}$$

where T is the temperature of the product stream, A is the frequency factor (= $4.711 \cdot 10^9$ s^{-1}), R is the gas constant, and E is the activation energy (= $7.536 \cdot 10^4$ J/mol). Plot the conversion from this material balance as a function of temperature of the product stream ($260 \le T \le 400$ K).

b. Plot on the same graph (created in part a) the conversion obtained from energy balance as a function of temperature of the product stream for three different feed temperatures (T_{in}): 263.15, 273.15, and 293.15 K. From energy balance, the conversion of reactant is given by:

$$X = \frac{C_{p,in}(T - T_{in})}{-\Delta H} \qquad (2)$$

where $C_{p,in}$ is the heat capacity of the reactant entering the reactor (= 1675 J/(mol·K)) and ΔH is the enthalpy change of reaction measured at the temperature of the product stream, which is obtained from:

$$\Delta H = \Delta H_R^0 + \Delta C_p(T - T_R)$$

where ΔC_p is the overall change in heat capacity per mol of reactant that is reacted (= –29.31 J/(mol·K))

c. For T_{in} = 273.15 K, determine the actual temperature of the product stream that gives the highest possible conversion. The actual temperature of the product stream and conversion can be obtained by simultaneously solving Equations (1) and (2). What is the actual conversion for this particular T_{in}?

d. Plot the actual temperature of the product stream that gives the highest possible conversion as a function of the feed temperature ($260 \le T_{in} \le 300$ K).

e. Plot the highest possible conversion as a function of the feed temperature ($260 \le T_{in} \le 300$ K). Do you think it is desirable to have a feed temperature colder than 270 K?

9. In manufacturing, a special coating on a curved solar absorber surface of area (A_2) 15 m² is cured by exposing it to an infrared heater of width (W) 1 m and area (A_1) 10 m². The absorber and heater are each of length (L) 10 m and are separated by a distance of (H) 1 m. The desired surface temperature of the heater is 1000 K (T_1). The heater has an emissivity of 0.9 (ε_1), while the absorber has an emissivity of 0.5 (ε_2). The system is in a large room, the wall temperature of which is 300 K (T_3). The needed net rate of heat transfer to the absorber surface (q_2) is 77.1 kW. The heat transfer is assumed to be entirely due to radiation.

a. Determine the heater power requirement (q_1). The heater power can be calculated from:

$$q_1 = \frac{\varepsilon_1 A_1}{1-\varepsilon_1}\left(\sigma T_1^4 - J_1\right)$$

where J_1 is the surface radiosity of the heater and σ is the Stefan-Boltzmann constant (= 5.67×10^{-8} W/(m²·K⁴). The surface radiosity of the heater can be calculated by solving the following equations:

$$\frac{\varepsilon_1 A_1}{1-\varepsilon_1}\left(\sigma T_1^4 - J_1\right) = F_{12}A_1\left(J_1 - J_2\right) + F_{13}A_1\left(J_1 - \sigma T_3^4\right)$$

$$\frac{\varepsilon_2 A_2}{1-\varepsilon_2}\left(\sigma T_2^4 - J_2\right) = F_{12}A_1\left(J_2 - J_1\right) + F_{23}A_2\left(J_2 - \sigma T_3^4\right)$$

$$q_2 = \frac{\varepsilon_2 A_2}{1-\varepsilon_2}\left(J_2 - \sigma T_2^4\right)$$

where J_2 is the surface radiosity of the absorber, T_2 is the surface temperature of the absorber, and F_{ij} is the view factor between surface i and surface j. These view factors are calculated as follows:

$$F_{12} = \frac{2}{\pi XY}\left\{\ln\left[\sqrt{\frac{\left(1+X^2\right)\left(1+Y^2\right)}{1+X^2+Y^2}}\right] + X\sqrt{1+Y^2}\tan^{-1}\left(\frac{X}{\sqrt{1+Y^2}}\right)\right.$$
$$\left. + Y\sqrt{1+X^2}\tan^{-1}\left(\frac{Y}{\sqrt{1+X^2}}\right) - X\tan^{-1}X - Y\tan^{-1}Y\right\}$$

$$\text{where } X = \frac{W}{H} \text{ , } Y = \frac{L}{H}$$

$$F_{13} = 1 - F_{12}$$

$$F_{23} = \frac{F_{13}A_1}{A_2}$$

b. To analyze the effect of the separation distance between the heater
   and the absorber on the heater power requirement, plot q_1 vs. H
   $(0.5 \leq H \leq 3$ m$)$, and on the surface temperature of the absorber,
   plot T_2 vs. $H (0.5 \leq H \leq 3$ m$)$. Discuss whether or not your plots are
   reasonable.

Chapter 5
Curve Fitting

Chemical engineers deal with a lot of data that are crucial for modeling, process analysis, and process design. Data are often obtained or given for discrete values along a continuum, from which we often require trend analysis and/or estimates at intermediate points. If we are developing a mathematical model for a certain process or physical phenomenon, we may also need to obtain some model parameters fitted to experimental data. The procedure that we need for these purposes is called curve fitting.

There are two general approaches for curve fitting, i.e., regression and interpolation. The choice of the approach is dependent on the extent of error associated with the data of interest and the purpose of the curve fitting. If the data exhibit a significant degree of error ("noise"), any individual data point may be incorrect. In this case, regression should be the correct approach, from which a trend or pattern of the data can be obtained. Interpolation should not be used for such data.

If the data are known to be very accurate, for example data originating from standard tables, we could use either regression or interpolation depending on our purpose. If we would like to obtain parameters of a model that we develop, we should use regression. On the other hand, if we only require estimates at intermediate points, interpolation is often the best approach.

5.1 Interpolation

Interpolation is the estimation of values between precise discrete points. We encounter this kind of estimation in many applications in chemical engineering, for example in the calculations of thermodynamic properties of water from the steam table.

5.1.1 Linear Interpolation

Linear interpolation is an interpolation using a linear function connecting two points. We use this linear interpolation when we know that the trend of the data is indeed linear or the interval between the two points is quite narrow.

In Mathcad, linear interpolation can be done easily using *linterp* function. When we use *linterp* function for a set of data, in fact we create linear functions, each of which connects two data points. Thus, if we have *n* data points, we generate *n-1* linear functions. This function supports units. Procedure 5.1 shows the use of this function.

5.1.2 Cubic Spline

If the trend of data is not linear and we want accurate estimates, cubic spline is often the best interpolation approach. In this approach, we create a third-order polynomial (cubic equation) for each interval between two adjacent data points. The cubic equation for interval *i* between two adjacent data points, i.e., (x_{i-1}, y_{i-1}) and (x_i, y_i), is as follows:

$$f_i(x) = \frac{f''(x_{i-1})}{6(x_i - x_{i-1})}(x_i - x)^3 + \frac{f''(x_i)}{6(x_i - x_{i-1})}(x - x_{i-1})^3$$

$$+ \left[\frac{y_{i-1}}{x_i - x_{i-1}} - \frac{f''(x_{i-1})(x_i - x_{i-1})}{6}\right](x_i - x) \qquad (5.1)$$

$$+ \left[\frac{y_i}{x_i - x_{i-1}} - \frac{f''(x_i)(x_i - x_{i-1})}{6}\right](x - x_{i-1})$$

where $f''(x_i)$ represents the second derivative of the function at $x = x_i$ and *i* runs from 1 to *N*, which is the total number of intervals ($N \geq 2$).

Unlike linear interpolation where the curve drawn over the whole range of independent variable is not smooth (the slope of the interpolating curves at the data points changes abruptly), the fitting curve generated using cubic spline is smooth. The first and second derivatives of the functions are thus continuous at the data points.

Cubic spline can be done in Mathcad using *lspline*, *pspline*, or *cspline* function to obtain the second derivatives at the data points, which are required in generating the fitting curve, as indicated in Equation (5.1). The *lspline*, *pspline*, and *cspline* functions differ in how the second derivatives at the end points, i.e., $f''(x_0)$ and $f''(x_N)$, are specified. The *lspline* function specifies that the second derivatives at the end points are zero, the *pspline* function specifies that the second derivatives at the end points are equal to those of their immediate neighbors, i.e., $f''(x_0) = f''(x_1)$ and $f''(x_N) = f''(x_{N-1})$, while *cspline* function specifies that the second derivative at x_0 is extrapolated linearly from $[x_1, f''(x_1)]$ and $[x_2, f''(x_2)]$ and that at x_N is extrapolated linearly from $[x_{N-1}, f''(x_{N-1})]$ and $[x_{N-2}, f''(x_{N-2})]$. In the case when $N = 2$, *cspline* function specifies that the second derivative at x_N is zero and that at x_0 is extrapolated linearly from

$[x_1, f'(x_1)]$ and $[x_2, f'(x_2)]$. After obtaining these derivatives, we use *interp* function to generate the fitting curve. These cubic spline functions and *interp* also support units. Procedure 5.2 describes the use of these functions.

These cubic spline functions and *interp* function can also be used to perform two-dimensional interpolation. The functions require that the data of the dependent variable be presented as a square matrix. In other words, the number of points of the first independent variable must be the same as that of the second independent variable. Procedure 5.3 demonstrates the applications of these functions for two-dimensional interpolation.

Procedure 5.1: *linterp*

For example, we have the following set of data consisting of an independent variable x and a dependent variable y:

x	1.5	3.0	4.0	5.5
y	5.2	7.8	9.2	11.1

We want to calculate the values of y at x = 2.0 and 5.0.

1. Construct two vectors for the independent and dependent variables:

$$vx := \begin{pmatrix} 1.5 \\ 3 \\ 4 \\ 5.5 \end{pmatrix} \qquad vy := \begin{pmatrix} 5.2 \\ 7.8 \\ 9.2 \\ 11.1 \end{pmatrix}$$

<u>Note</u>: The elements in the vector of the independent variable must be in ascending order

2. Plot these data to see if linear interpolation is a suitable approach.

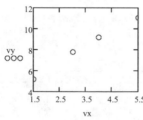

3. Create a function to represent the interpolating functions (fitting curve) for all intervals by using *linterp* function:

$$y(x) := linterp(vx, vy, x)$$

Although we can directly use *linterp* function to calculate y at any x without creating a function, it is a good practice to create such a function.

4. Calculate the y values:

at x = 2: $\quad y(2) = 6.067$

at x = 5: $\quad y(5) = 10.467$

For extrapolation, i.e., x is outside the data range, one of the two functions of the end intervals will be used depending on which side of the data range the extrapolation is performed.

5. We could also plot the fitting curve:

Each interval is represented by a linear function connecting the two end points of the interval.

Procedure 5.2: _lspline/pspline/cspline & interp_ (One dimensional interpolation)

For example, we have the following set of data consisting of an independent variable x and a dependent variable y and we want to calculate the values of y at x = 5.0 and 10.0:

x	2.0	3.0	4.0	5.5	7.0	9.0	11.0
y	5.2	7.9	9.5	11.0	11.8	12.7	14.5

1. Construct two vectors for x and y:

Note: The elements in the vector of the independent variable must be in ascending order

$$vx := \begin{pmatrix} 2.0 \\ 3.0 \\ 4.0 \\ 5.5 \\ 7.0 \\ 9.0 \\ 11.0 \end{pmatrix} \qquad vy := \begin{pmatrix} 5.2 \\ 7.9 \\ 9.5 \\ 11.0 \\ 11.8 \\ 12.7 \\ 14.5 \end{pmatrix}$$

2. Calculate the second derivatives needed using _lspline/pspline/cspline_ function:

$$vs := cspline(vx, vy) \qquad \text{The syntax for } \textit{lspline} \text{ or } \textit{pspline} \text{ is the same.}$$

3. Create a function to represent the fitting curve using _interp_ function:

$$y(x) := interp(vs, vx, vy, x)$$

4. Calculate the y values: $y(5) = 10.593$ $y(10) = 13.416$

5. It is always a good idea to plot the fitting curve:

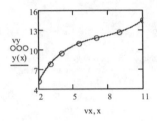

In this case, each interval is represented by a cubic equation connecting the two end points of the interval.

Extrapoation to x values beyond the data range is not recommended. If we extrapolate, one of the two functions of the end intervals will be used depending on which side of the data range the extrapolation is performed.

Procedure 5.3: _lspline/pspline/cspline & interp_ (Two dimensional interpolation)

For example, we have the following set of z data that depend on two independent variables x and y and we want to calculate the value of z at x = 2.5 and y = 7.0:

x\y	5.2	7.9	9.5	11.0
2.0	11.173	12.209	12.714	13.139
3.0	12.126	13.163	13.668	14.092
4.0	12.930	13.967	14.472	14.896
5.5	13.966	15.002	15.507	15.932

1. Construct two vectors for x and y:

Note: The elements in the vectors of the independent variables must be in ascending order

$$vx := \begin{pmatrix} 2.0 \\ 3.0 \\ 4.0 \\ 5.5 \end{pmatrix} \qquad vy := \begin{pmatrix} 5.2 \\ 7.9 \\ 9.5 \\ 11.0 \end{pmatrix}$$

2. Construct a matrix augmenting these two vectors: $Mxy := augment(vx, vy)$

3. Construct a matrix containing the z data:

$$Mz := \begin{pmatrix} 11.173 & 12.209 & 12.714 & 13.139 \\ 12.126 & 13.163 & 13.668 & 14.092 \\ 12.93 & 13.967 & 14.472 & 14.896 \\ 13.966 & 15.002 & 15.507 & 15.932 \end{pmatrix}$$

4. Calculate the second derivatives needed using _cspline_ function:

$$vs := cspline(Mxy, Mz) \qquad \text{The first argument is the matrix containing the independent variables and the second is the matrix containing the dependent variables.}$$

5. Create a function to represent the fitting curve using *interp* function:

$$z(x, y) := interp\left[vs, Mxy, Mz, \begin{pmatrix} x \\ y \end{pmatrix}\right]$$

6. Calculate the z value: $z(2.5, 7.0) = 12.392$

Example set 5.1

1. The mean molar heat capacity of nitrogen between 298 K and T at 1 atm has been tabulated as follow:

T [K]	373.0	473.0	573.0	673.0	773.0	873.0	973.0
c_p [J/mol.K]	29.19	29.29	29.46	29.68	29.97	30.27	30.56

Determine the mean molar heat capacity of nitrogen between 298 K and
a. 450 K
b. 525 K
c. 740 K

Solution:

$$T := \begin{pmatrix} 373 \\ 473 \\ 573 \\ 673 \\ 773 \\ 873 \\ 973 \end{pmatrix} \quad c_p := \begin{pmatrix} 29.19 \\ 29.29 \\ 29.46 \\ 29.68 \\ 29.97 \\ 30.27 \\ 30.56 \end{pmatrix} \quad y(x) := linterp(T, c_p, x)$$

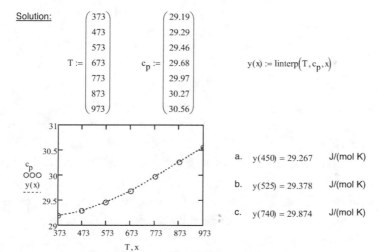

a. $y(450) = 29.267$ J/(mol K)

b. $y(525) = 29.378$ J/(mol K)

c. $y(740) = 29.874$ J/(mol K)

2. The vapor pressure (P) of water and the specific volume (v) of saturated water vapor are tabulated in a steam table:

t [°C]	0.01	3.0	6.0	9.0	12.0	15.0	18.0
P [kPa]	0.6113	0.7577	0.9349	1.1477	1.4022	1.7051	2.0640
v [m³/kg]	206.136	168.132	137.734	113.386	93.784	77.926	65.038

Determine the vapor pressure of water and the specific volume (v) of saturated water vapor at:
a. 1°C
b. 10°C
c. 14°C
using cubic spline.

Solution:

$$t := \begin{pmatrix} 0.01 \\ 3 \\ 6 \\ 9 \\ 12 \\ 15 \\ 18 \end{pmatrix} °C \quad P := \begin{pmatrix} 0.6113 \\ 0.7577 \\ 0.9349 \\ 1.1477 \\ 1.4022 \\ 1.7051 \\ 2.0640 \end{pmatrix} \cdot kPa \quad v := \begin{pmatrix} 206.136 \\ 168.132 \\ 137.734 \\ 113.386 \\ 93.784 \\ 77.926 \\ 65.038 \end{pmatrix} \cdot \frac{m^3}{kg}$$

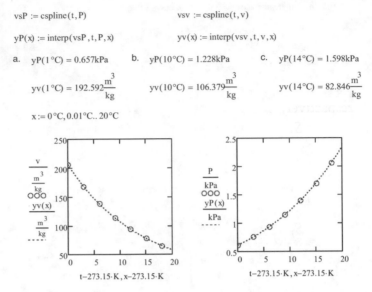

$$vsP := cspline(t, P) \qquad\qquad vsv := cspline(t, v)$$

$$yP(x) := interp(vsP, t, P, x) \qquad\qquad yv(x) := interp(vsv, t, v, x)$$

a. $yP(1\,°C) = 0.657\,kPa$ b. $yP(10\,°C) = 1.228\,kPa$ c. $yP(14\,°C) = 1.598\,kPa$

$$yv(1\,°C) = 192.592\frac{m^3}{kg} \qquad yv(10\,°C) = 106.379\frac{m^3}{kg} \qquad yv(14\,°C) = 82.846\frac{m^3}{kg}$$

$$x := 0\,°C, 0.01\,°C .. 20\,°C$$

5.2 Regression

In regression, an overall trend is characterized by specifying a function beforehand. If a theoretical model is not available, visual inspection of the data plot is usually helpful in determining a trial function. If we can transform a non-linear model into a linear model, linear regression is often preferable. For example, consider a set of data, C_R and t, which will be characterized by the following model:

$$\frac{1}{C_R} - \frac{1}{C_{Ro}} = k \cdot t$$

where C_{Ro} and k are constants. This model can be considered as a non-linear model if we plot C_R vs. t. However, we can transform this non-linear model into a linear model by defining a new dependent variable y, which is equal to $1/C_R$. Thus, the model becomes:

$$y = k \cdot t + \frac{1}{C_{Ro}}$$

which is obviously a linear model.

After we obtain the parameters of the model, some criteria are then devised to quantify the adequacy of the fit. The most widely used criterion is the correlation coefficient (r) or the coefficient of determination (r^2). For a set of data consisting of independent variable x and dependent variable y, which is modeled with $y = f(x)$, the correlation coefficient and the coefficient of determination are given by

$$r = \sqrt{\frac{S_t - S_r}{S_t}}$$ (5.2)

$$r^2 = \frac{S_t - S_r}{S_t}$$ (5.3)

respectively, where

S_t = the sum of the squares around the mean of the dependent variable y (the spread of the dependent variable that exists prior to regression)

$$= \sum_i (y_i - \bar{y})^2$$

S_r = the sum of the squares of the residuals around the fitting curve (the spread that remains after regression)

$$= \sum_i [y_i - f(x_i)]^2$$

$\bar{y}$ = the mean of the data

y_i = the individual data of the dependent variable

x_i = the individual data of the independent variable

If a model fits the data perfectly, $S_r = 0$ and thus the correlation coefficient or the coefficient of determination is equal to unity. Although the correlation coefficient provides a measure of the degree of the adequacy of the fit, we have to be careful in interpreting its value. Just because r is close to 1 does not guarantee that the fit or the model is correct.

5.2.1 Linear Regression

In Mathcad, linear regression can be done using several built-in functions, i.e., *intercept* and *slope* functions, *line* function, and *regress* function. Since *regress* function is a general function for polynomial regression, its procedure will be described in the next section. Procedures 5.4 and 5.5 show how to use *intercept/slope* and *line* functions, respectively.

Slope and *intercept* functions support units, but not *line* function because the result from this function is a vector, the element of which cannot have different units.

To measure the goodness-of-fit, we calculate the correlation coefficient using *corr* function, which corresponds to Equation (5.2). The *corr* function supports units. We show the use of *corr* function in Procedure 5.4.

Procedure 5.4: _intercept & slope_

For example, we have the following data:

x	1.0	2.2	3.5	4.3	5.0
y	1.8	4.5	7.2	8.4	9.7

and we want to regress them using a linear model: $y = a_0 + a_1 \cdot x$

1. Create two vectors representing independent variable x and dependent variable y:

$$vx := \begin{pmatrix} 1.0 \\ 2.2 \\ 3.5 \\ 4.3 \\ 5.0 \end{pmatrix} \qquad vy := \begin{pmatrix} 1.8 \\ 4.5 \\ 6.7 \\ 8.4 \\ 10.2 \end{pmatrix}$$

2. Plot the data to visually inspect if a linear model would do the job:

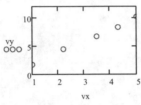

3. Calculate the intercept (a_0) and the slope (a_1) of the fitting curve using *intercept* and *slope* functions:

$$a_0 := \text{intercept}(vx, vy) = -0.2 \qquad a_1 := \text{slope}(vx, vy) = 2.038$$

4. Create a function representing the model, i.e., the fitting curve:

$$y(x) := a_0 + a_1 \cdot x \qquad \text{y(x) is a scalar function and its argument (x) is also a scalar.}$$

5. Plot this fitting curve on the data plot to visually check the model:

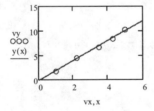

It is always a good idea to visually check the goodness-of-fit of the model.

Note: vx is a vector and x is a scalar.

$$r := \text{corr}\left(vy, \overrightarrow{y(vx)}\right) = 0.998$$

Thus, the dependent variable vy is well correlated/fitted by the model y(x).

Procedure 5.5: _line_

For the example of Procedure 5.4, we can also use *line* function to obtain the parameters (intercept and slope) of the fitting curve. Only number 3 of the above procedure is changed:

3. Calculate the slope (a_1) and the intercept (a_0) of the fitting curve using *line* function:

$$a := \text{line}(vx, vy)$$

$$a = \begin{pmatrix} -0.2 \\ 2.038 \end{pmatrix}$$

Note that *line* function generate a vector containing the coefficients (intercept and slope) of the line.

$$a_0 = -0.2$$

$$a_1 = 2.038$$

5.2.2 Polynomial Regression

For polynomial regression, we use *regress* function to obtain the parameters of the fitting curve and use *interp* function or write our own polynomial to generate the fitting curve. Of course, we can use these functions for linear regression; a linear function is a first order polynomial. The adequacy of the fit is then also measured using the correlation coefficient *r*. As we can expect, *regress* function does not support units. Procedure 5.6 describes the use of these functions.

Procedure 5.6: *regress & interp*

For example, we have the following data:

x	1.0	2.0	3.5	4.0	5.0
y	1.2	4.7	12.0	16.5	24.7

and we want to regress them using 2nd order polynomial: $y = a_0 + a_1 \cdot x + a_2 \cdot x^2$

1. Create two vectors representing independent variable x and dependent variable y:

2. Plot the data to visually inspect if the model would do the job. The plot is combined in step 5.

$$vx := \begin{pmatrix} 1.0 \\ 2.0 \\ 3.5 \\ 4.0 \\ 5.0 \end{pmatrix} \quad vy := \begin{pmatrix} 1.2 \\ 4.7 \\ 12.0 \\ 16.5 \\ 24.7 \end{pmatrix}$$

3. Calculate the parameters (a_0, a_1, and a_2) of the fitting curve using *regress* function:

The order of polynomial: $k := 2$ $v := \text{regress}(vx, vy, k)$

$$v = \begin{pmatrix} 3 \\ 3 \\ 2 \\ 0.017 \\ 0.377 \\ 0.913 \end{pmatrix}$$

The first three members of this vector are needed by MathCad. The rest are the coefficients of the polynomial we need.

$i := 0 .. k$

$a_i := v_{i+3}$

Note that all the subscripts used in this example are indices of arrays, not labels (literal subscripts). This way, we can simplify the assignments.

4. Create a function representing the model, i.e., the fitting curve, using *interp* function:

$y(x) := \text{interp}(v, vx, vy, x)$ x is a scalar

or create our own polynomial: $y(x) := \sum_{j=0}^{k} \left(a_j \cdot x^j \right)$ k-th order polynomial

5. Calculate the correlation coefficient and plot the fitting curve on the data plot to visually check the model:

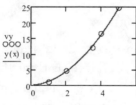

$$r := \text{corr}\left(vy, \overrightarrow{y(vx)}\right) = 0.999$$

These functions can also be used to perform multivariate regression, as demonstrated in Procedure 5.7.

Procedure 5.7: *regress & interp*(Multivariate regression)

Suppose that z is a function of x and y, and the following experimental data are obtained:

x	2	3	4	5	6	2.5	8	4.7
y	3	5	7	9	11	10.5	15	10.3
z	7.90	8.45	18.09	33.03	45.20	25.60	71.05	32.50

We want to regress the data using a second order polynomial with 2 independent variables:

$$z(x, y) = \sum_{i=0}^{2} \sum_{j=0}^{2-i} \left(c_{i,j} \cdot x^i \cdot y^j \right) \quad \text{and calculate z at x = 7.5 and y = 12.}$$

1. Create 3 vectors containing the values of x, y, and z:

$$
vx := \begin{pmatrix} 2 \\ 3 \\ 4 \\ 5 \\ 6 \\ 2.5 \\ 8 \\ 4.7 \end{pmatrix} \qquad vy := \begin{pmatrix} 3 \\ 5 \\ 7 \\ 9 \\ 11 \\ 10.5 \\ 15 \\ 10.3 \end{pmatrix} \qquad vz := \begin{pmatrix} 7.90 \\ 8.45 \\ 18.09 \\ 33.03 \\ 45.20 \\ 25.60 \\ 71.05 \\ 32.50 \end{pmatrix}
$$

2. Create a matrix by augmenting vectors containing independent variables:

$$
Mxy := augment(vx, vy) \qquad Mxy^T = \begin{pmatrix} 2 & 3 & 4 & 5 & 6 & 2.5 & 8 & 4.7 \\ 3 & 5 & 7 & 9 & 11 & 10.5 & 15 & 10.3 \end{pmatrix}
$$

3. Use *regress* function to obtain the parameters of the fitting surface:

$$
vs := regress(Mxy, vz, 2) \qquad vs^T = (3 \; 3 \; 2 \; 0.202 \; -0.125 \; 3.958 \; 0.713 \; -5.694 \; 0.961)
$$

4. Create a function representing the fitting surface by using *interp* function and calculate z:

$$
z(x, y) := interp\left[vs, Mxy, vz, \begin{pmatrix} x \\ y \end{pmatrix} \right] \qquad z(7.5, 12) = 59.787
$$

5. Plot the data and fitting surface:

This surface plot can be created from the Graph Toolbar. To plot the data points, type vectors vx, vy, and vz (in a parenthesis) in the empty place holder. To plot the fitting surface, type a comma followed by the function name. The ranges of x and y, in which the fitting surface z is created, have been modified (double click the figure, choose Quick Plot tab, and change the start and end values).

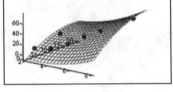

$(vx, vy, vz), z$

5.2.3 Non-Linear Regression

Mathcad provides several built-in functions for several non-linear models, such as exponential or logarithmic function. Here, we will discuss only *genfit* function, which is a general function for any non-linear model. Since the result from *genfit* function is a vector containing the model parameters that might have different units, this function does not completely support units. Procedure 5.8 shows the use of this function.

Similar to *find* function for solving a system of nonlinear equations, *genfit* function is also highly sensitive to initial guesses. If *genfit* function does not converge, one may try to change the numerical method used by right clicking on the *genfit* function and choosing the desired method, to change the initial guesses, or to scale the data so that all model parameters are of similar order of magnitude. Plotting the fitting function with the guess values could also help us in refining the guess values.

Procedure 5.8: *genfit*

For example, we have the following data:

x	1.1	1.2	1.4	1.5	1.6	1.8	2.0	2.1
y	3.1	2.7	1.8	1.7	1.4	1.3	0.7	0.6

and we want to regress them using a non-linear model: $y = \dfrac{A \cdot e^{-x}}{1 + B \cdot x}$

1. Create two vectors representing independent variable x and dependent variable y:

2. Create a vector containing the initial guesses of A and B and a function representing the model function:

$$vx := \begin{pmatrix} 1.1 \\ 1.2 \\ 1.4 \\ 1.5 \\ 1.6 \\ 1.8 \\ 2.0 \\ 2.1 \end{pmatrix} \qquad vy := \begin{pmatrix} 3.1 \\ 2.7 \\ 1.8 \\ 1.7 \\ 1.4 \\ 1.3 \\ 0.7 \\ 0.6 \end{pmatrix}$$

$$vg := \begin{pmatrix} 1 \\ 1 \end{pmatrix}$$ The first element is the initial guess for A and the second is for B.

$$y(x, A, B) := \frac{A \cdot e^{-x}}{1 + B \cdot x}$$ Note that the parameters must be listed after the independent x.

3. Calculate the parameters (A and B) of the fitting curve using *genfit* function:

$$\begin{pmatrix} A \\ B \end{pmatrix} := \text{genfit}(vx, vy, vg, y) = \begin{pmatrix} 33.226 \\ 2.319 \end{pmatrix}$$

4. Calculate the correlation coefficient and plot the fitting curve on the data plot to visually check the model:

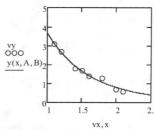

vy
○○○
y(x, A, B)

vx, x

$$r := \text{corr}\left(vy, \overrightarrow{y(vx, A, B)}\right)$$

$$r = 0.991$$

Example set 5.2

1. Flowmeter calibration data are as follows:

Reading, X	5	20	40	70	80
Flow rate, F [liters/hour]	130.2	500.3	1050.5	1653.2	2013.7

What is the flow rate if the flowmeter reading is 53?

Solution:

Create vectors containing the given data:

$$X := \begin{pmatrix} 5 \\ 20 \\ 40 \\ 70 \\ 80 \end{pmatrix} \qquad F := \begin{pmatrix} 130.2 \\ 500.3 \\ 1050.5 \\ 1653.2 \\ 2013.7 \end{pmatrix}$$

Plot the data to see the relationship between X and F:

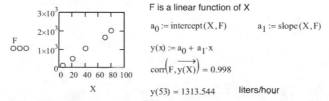

F
○○○

X

F is a linear function of X

$a_0 := \text{intercept}(X, F) \qquad a_1 := \text{slope}(X, F)$

$$y(x) := a_0 + a_1 \cdot x$$

$$\text{corr}\left(F, \overrightarrow{y(X)}\right) = 0.998$$

$$y(53) = 1313.544 \qquad \text{liters/hour}$$

Plot the fitting curve to visually check the goodness-of-fit

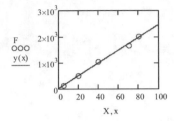

$\begin{array}{c} F \\ \circ\circ\circ \\ \dfrac{y(x)}{} \end{array}$

X, x

2. Experimental density (ρ) data as a function of pressure (P) for N_2 at 200 K (T) are as follow:

P [atm]	3.2	6.0	9.0	12.0	14.0	17.0	19.0	21.0
ρ [mol/liter]	0.1995	0.3825	0.5895	0.8108	0.9685	1.2257	1.4159	1.6281

Using these data, estimate the second, third, and fourth virial coefficients.

The virial equation of state that relates pressure and density is:

$$Z = \frac{P}{\rho \cdot R \cdot T} = 1 + B \cdot \rho + C \cdot \rho^2 + D \cdot \rho^3 \qquad R = \text{gas constant} = 0.08206 \; \frac{\text{liter} \cdot \text{atm}}{\text{mol} \cdot \text{K}}$$

where B, C, and D are the second, third, and fourth virial coefficients, respectively.

Solution:

The virial equation of state is transformed into:

$$\frac{Z-1}{\rho} = B + C \cdot \rho + D \cdot \rho^2 \qquad \text{second order polynomial} \qquad k := 2$$

$R := 0.08206 \qquad T := 200$

$$\rho := \begin{pmatrix} 0.1995 \\ 0.3825 \\ 0.5895 \\ 0.8108 \\ 0.9685 \\ 1.2257 \\ 1.4159 \\ 1.6281 \end{pmatrix} \qquad P := \begin{pmatrix} 3.2 \\ 6.0 \\ 9.0 \\ 12.0 \\ 14.0 \\ 17.0 \\ 19.0 \\ 21.0 \end{pmatrix} \qquad \overrightarrow{Z := \frac{P}{\rho \cdot R \cdot T}} \qquad \overrightarrow{Y := \frac{Z-1}{\rho}}$$

Note that vectorization is used to calculate element by element in a quick way.

Calculate the coefficients: $v := \text{regress}(\rho, Y, k)$

$i := 0..k \qquad a_i := v_{i+3}$

$B := a_0 = -0.111$

$C := a_1 = -0.012$

$D := a_2 = -2.08 \times 10^{-4}$

Plot the data and the fitting curve to check the goodness-of fit:

$y(x) := \text{interp}(v, \rho, Y, x)$

$\begin{array}{c} Y \\ \circ\circ\circ \\ \dfrac{y(x)}{} \end{array}$

ρ, x

The correlation coefficient:

$$r := \text{corr}\left(Y, \overrightarrow{y(\rho)}\right) = 1$$

3. A slab of wood 50.8 10^{-3} m thick (2h) is dried from both sides by air. The following data are obtained from the experiment:

Time, t [hours]	5	10	15	20	25	30
Wood free moisture, X [kg H_2O/kg dry wood]	0.245	0.230	0.211	0.197	0.187	0.176

The free moisture in wood as a function of time can be approximated by

$$X(t) = X_0 \cdot \frac{8}{\pi^2} \left[e^{-D \cdot t \cdot \left(\frac{\pi}{2 \cdot h}\right)^2} + \frac{1}{9} \cdot e^{-9 \cdot D \cdot t \cdot \left(\frac{\pi}{2 \cdot h}\right)^2} \right]$$

where X_0 is the initial free moisture of the wood [kg H_2O/kg dry wood], h is half of the wood thickness [m], and D is the difussivity of water in the wood [m^2/hour].

Estimate the difussivity of water in the wood.

<u>Solution:</u> $h := 25.4 \, 10^{-3}$

Two parameters, i.e., X_0 and D, are to be fitted.

$$vt := \begin{pmatrix} 5 \\ 10 \\ 15 \\ 20 \\ 25 \\ 30 \end{pmatrix} \quad vX := \begin{pmatrix} 0.245 \\ 0.230 \\ 0.211 \\ 0.197 \\ 0.187 \\ 0.176 \end{pmatrix} \quad F(t, X_0, D) := X_0 \cdot \frac{8}{\pi^2} \left[e^{-D \cdot t \cdot \left(\frac{\pi}{2 \cdot h}\right)^2} + \frac{1}{9} \cdot e^{-9 \cdot D \cdot t \cdot \left(\frac{\pi}{2 \cdot h}\right)^2} \right]$$

Initial guesses: $G := \begin{pmatrix} 0.27 \\ 1 \cdot 10^{-6} \end{pmatrix}$

$$\begin{pmatrix} X_0 \\ D \end{pmatrix} := genfit(vt, vX, G, F) = \begin{pmatrix} 0.299 \\ 2.879 \times 10^{-6} \end{pmatrix}$$

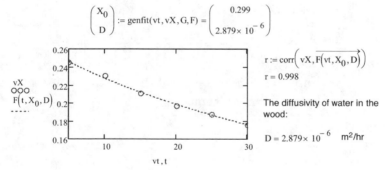

$r := corr\left(vX, \overrightarrow{F(vt, X_0, D)}\right)$

$r = 0.998$

The diffusivity of water in the wood:

$D = 2.879 \times 10^{-6}$ m^2/hr

Problems

1. Accurate data on the liquid molar volume (V_L) of nitrogen at melting points are as follows:

T [K]	64.0	68.0	72.0	76.0	80.0
V_L [cc/mol]	32.225	31.912	31.527	31.124	30.731

Estimate the liquid molar volume of nitrogen at a melting point of 70 K.

2. The mean activity coefficients (γ) of aqueous solutions of Na_2SO_4 have been tabulated as follows:

m	0.1	0.3	0.5	0.7	0.9	1	1.4
γ	0.4483	0.3230	0.2692	0.2360	0.2128	0.2036	0.1769

m	1.8	2.2	2.6	3	3.4	3.8
γ	0.1604	0.1498	0.1431	0.1394	0.1380	0.1386

What are the activity coefficients of aqueous solutions of Na_2SO_4 for molality (m) of 0.2 and 1.2? Also determine the molality of the solution that has an activity coefficient of 0.4. Use cubic spline.

3. The enthalpy of superheated steam at a certain condition needs to be accurately determined. From the steam table, the following enthalpy [H, kJ/kg] data are available:

P [kPa]\T [K]	300	350	400	450	500
150	3073.3	3174.7	3277.5	3381.7	3487.6
200	3072.1	3173.8	3276.7	3381.1	3487.0
250	3070.9	3172.8	3275.9	3380.4	3486.5
300	3069.7	3171.9	3275.2	3379.8	3486.0

Determine the enthalpy of superheated steam at 420 K and 190 kPa.

4. Experimental data of reaction constant (k) for a first order reaction as a function of temperature (T) are as follows:

T [K]	303.0	310.0	314.0	319.0	325.0
k [s^{-1}]	0.0053	0.0113	0.0190	0.0365	0.0727

The temperature dependence of the reaction constant could be correlated by the *Arrhenius* equation:

$$k = Ae^{-\frac{E}{RT}}$$

where A is the frequency factor, R is the gas constant (=8.314 J/mol.K), and E is the activation energy. Determine the activation energy for this reaction. What is the reaction constant at 305 K?

5. A reactive substance R reacts according to the following reaction:

$$R \rightarrow \text{products}$$

The following data are obtained from experiment:

Time, t [sec]	0	2	5	8	10	20
Concentration of R, C_R [mol/liter]	1.5	1.0	0.7	0.6	0.5	0.3

We would like to analyze the experimental data and suggest what the order of reaction is. Then, we also want to determine the reaction constant k.

For a zeroth order reaction, the following relation must be satisfied:

$$C_{R0} - C_R = kt$$

where C_{R0} is the initial concentration of R (C_R at $t = 0$) and k is the reaction constant.

For a first order reaction, the following relation must be satisfied:

$$\ln(C_{R0}) - \ln(C_R) = kt$$

For a second order reaction, the following relation must be satisfied:

$$\frac{1}{C_R} - \frac{1}{C_{R0}} = kt$$

Suggest the most probable order of reaction for this reaction (explain the reason) and then determine the reaction constant. What is the unit of k for this order of reaction? Estimate the concentration of the reactant after 30 seconds.

6. Experimental data of relative vapor pressure of glycerol-water mixtures at 20°C are as follows:

Weight fraction of glycerol	0.000	0.200	0.250	0.350	0.356	0.479	0.500
Relative vapor pressure	1.000	0.942	0.923	0.887	0.885	0.825	0.814

Weight fraction of glycerol	0.600	0.648	0.750	0.786	0.830	0.920	1.000
Relative vapor pressure	0.737	0.695	0.587	0.535	0.446	0.275	0.000

Perform a polynomial regression to obtain a model for relative vapor pressure as a function of weight fraction of glycerol. Then, estimate the relative vapor pressure for a mixture containing 10 weight% glycerol. Note: You need to investigate the order of polynomial that best fits the data.

7. Experimental vapor pressure (P_{sat}) data as a function of temperature (t) for chloroform are as follows:

t [°C]	0	10	20	30	40	50	60	70
P_{sat} [mmHg]	61.5	101.0	159.3	242.6	357.9	513.6	718.7	983.2

t [°C]	80	90	100	110	120	130	140	150
P_{sat} [mmHg]	1317.7	1733.9	2243.7	2859.2	3593.0	4457.9	5466.4	6631.0

a. Using the experimental data above, determine the parameters of Antoine equation for chloroform. Antoine equation relates vapor pressure to temperature.

Antoine equation: $P_{sat} = \exp\left(A - \dfrac{B}{t+C}\right)$

where A, B, and C are the Antoine parameters.

b. Using the parameters obtained in part (a) and Antoine equation, estimate the vapor pressure of chloroform at 56°C.

8. In the design of an absorption column (packed column), the mole fractions of solute at the gas-liquid interface (x_i, y_i) at several points in the column must be determined. x_i is the mole fraction of solute at the interface on the liquid side and y_i is the mole fraction of solute at the interface on the gas side. From the concept of mass transfer, it is known that:

$$\frac{k'_y a}{(1-y)_{iM}}(y - y_i) = \frac{k'_x a}{(1-x)_{iM}}(x_i - x)$$

from which x_i and y_i will be calculated. In the above equation, k_y' is the gas-film mass transfer coefficient, k_x' is the liquid-film mass transfer coefficient, a is the interfacial area per unit volume of packing, y and x are the mole fractions of solute in the bulk gas and liquid phases, respectively, at any point in the column, and the terms in the denominator are the log-mean averages given by the following expressions:

$$(1-y)_{iM} = \frac{(1-y)-(1-y_i)}{\ln\left(\dfrac{1-y}{1-y_i}\right)} \qquad (1-x)_{iM} = \frac{(1-x)-(1-x_i)}{\ln\left(\dfrac{1-x}{1-x_i}\right)}$$

At a certain condition, y and x are related to each other through an operating curve, which is derived from material balance:

$$y = \frac{\dfrac{L'}{V'}\dfrac{x}{1-x} + \left(\dfrac{y_b}{1-y_b} - \dfrac{L'}{V'}\dfrac{x_b}{1-x_b}\right)}{1 + \dfrac{L'}{V'}\dfrac{x}{1-x} + \left(\dfrac{y_b}{1-y_b} - \dfrac{L'}{V'}\dfrac{x_b}{1-x_b}\right)}$$

where L' is the molar flow rate of solute-free liquid (= 2800 lbmol/hr), V' is the molar flow rate of solute-free gas (= 900 lbmol/hr), and y_b (= 0.4) and x_b (= 0.16) are the mole fractions of solute in the bulk gas and liquid phases at the bottom of the column, respectively. Since the mass transfer resistance at the interface can usually be neglected, y_i and x_i are related to each other through an equilibrium curve (equilibrium relationship):

$$y_i = f_{eq}(x_i)$$

For this system and the given conditions, the following equilibrium data are available:

x_i	0	0.0126	0.0167	0.0208	0.0258	0.0309	0.0405
y_i	0	0.0151	0.0201	0.0254	0.0321	0.0390	0.0527

x_i	0.0503	0.0737	0.096	0.137	0.175	0.210
y_i	0.0671	0.105	0.145	0.235	0.342	0.463

a. From the given equilibrium data, use a curve fitting method to create the equilibrium function $f_{eq}(x_i)$.

b. Determine the mole fractions of solute at the gas-liquid interface (x_i, y_i) at several points in the column where the mole fractions of solute in the bulk liquid phase are x = 0, 0.02, 0.04, 0.06, 0.08, 0.10, 0.12, 0.14, and 0.16

Data: $k_y'a = 30$ lbmol/(hr·ft³) and $k_x'a = 250$ lbmol/(hr·ft³)

9. The activity coefficients of chloroform and 1,4 dioxane in a binary liquid mixture of chloroform (1)/1,4 dioxane (2) at 50°C can be estimated from Margules equations:

$$\gamma_1 = \exp\{x_2^2(A_{12} + 2(A_{21} - A_{12})x_1)\}$$
$$\gamma_2 = \exp\{x_1^2(A_{21} + 2(A_{12} - A_{21})x_2)\}$$

where γ_1 is the activity coefficient of chloroform, γ_2 is the activity coefficient of 1,4 dioxane, x_1 is the mol fraction of chloroform, x_2 is the mol fraction of 1,4 dioxane ($x_2 = 1 - x_1$), and A_{12} and A_{21} are the Margules parameters for this binary system. These two parameters are usually obtained from vapor-liquid equilibrium data. Determine the mol fraction of chloroform in the mixture such that $\gamma_1 = \gamma_2$.

From vapor-liquid equilibrium experiment, the following data are obtained:

x_1	0.0932	0.1248	0.1757	0.2000	0.2626
G^E/RT	−0.064	−0.086	−0.120	−0.133	−0.171

x_1	0.3615	0.4750	0.5555	0.6718
G^E/RT	−0.212	−0.248	−0.252	−0.245

where G^E/RT is the dimensionless excess Gibbs free energy. To obtain A_{12} and A_{21}, the following relation can be applied:

$$\frac{G^E}{RTx_1x_2} = A_{21}x_1 + A_{12}x_2$$

Chapter 6
Differentiation and Integration

In chemical engineering, differentiation and integration are often performed. If the function is known, we may try to differentiate or integrate symbolically using the derivative or integral operator with the symbolic equal sign. If the function is too complicated, we could use the derivative or integral operator and obtain the derivatives or integrals numerically. These operators can be accessed from the Calculus Toolbar in the Math Toolbar.

In some cases, we need to find the derivatives or integrals from a set of data. In this case, we have to find the function(s) that can well represent the data. This can be done by using a method described in Chapter 5. We use regression when the error of the experimental data is big. When the data are smooth, we could use either regression or interpolation; interpolation is usually the easiest approach when we do not know the type of the function.

6.1 Differentiation

Procedure 6.1 shows the symbolic (analytical) differentiation of known function, Procedure 6.2 shows the numerical differentiation of known function using a derivative operator, and Procedure 6.3 shows the differentiation from a set of data.

Procedure 6.1: *Symbolic Differentiation*

For example we need to calculate the first derivative of the following function at x = 0.1:

$$f(x) := x^3 + 3 \cdot x + \sin(x) + 3 \cdot \sqrt{x} \cdot \exp\left(2 \cdot x^2\right)$$

1. Use the derivative operator from the Calculus Toolbox and the symbolic equal sign and assign the result to a function, for example f'(x):

$$f'(x) := \frac{d}{dx} f(x) \rightarrow 3 \cdot x^2 + 3 + \cos(x) + \frac{3}{2 \cdot x^{\frac{1}{2}}} \cdot e^{2 \cdot x^2} + 12 \cdot x^{\frac{3}{2}} \cdot e^{2 \cdot x^2}$$

Note: The prime (') symbol is typed using [Ctrl][F7].

2. To calculate the derivative of f(x) at x = 0.1: $f'(0.1) = 9.251$

Procedure 6.2: *Numerical Differentiation*

For example we need to calculate the second derivative with respect to x of the following function at x = 0.1 and y = 0.2:

$$f(x, y) := \frac{x^3 \cdot y + 3 \cdot x + \sin(x) + 3 \cdot \sqrt{x \cdot y} \cdot \exp\left(2 \cdot x^2\right)}{\left(y + x^2 \cdot \tan(x)\right) \cdot \ln(x) - x \cdot y}$$

1. Use the derivative operator from the Calculus Toolbox and directly assign to a new function:

$$f_{xx}(x,y) := \frac{\partial^2}{\partial x^2} f(x,y)$$

Note: The default derivative operator can be changed to partial derivative operator by right clicking the mouse on the derivative operator and clicking View Derivative As Partial Derivative.

2. Calculate the derivative at x = 0.1 and y = 0.2 using the above function:

$$f_{xx}(0.1, 0.2) = -49.781$$

Procedure 6.3: _Differentiation of Data_

For example we need to determine the first derivative of y at x = 4.8 using the following data:

$$x := \begin{pmatrix} 1 \\ 1.8 \\ 2.5 \\ 3.2 \\ 4.1 \\ 5.4 \end{pmatrix} \qquad y := \begin{pmatrix} 2.5 \\ 5.1 \\ 11.0 \\ 18.5 \\ 30.0 \\ 50.7 \end{pmatrix}$$

1. Plot y vs. x to observe the smoothness of the data. From the plot below, we know that the data are smooth. Thus, we could use cubic spline interpolation to fit the data.

$$vs := cspline(x,y)$$

$$y1(x1) := interp(vs, x, y, x1)$$

2. Plot the fitting curve to check the interpolation

3. Calculate the derivative at x1 = 4.8:

$$y1'(x1) := \frac{d}{dx1} y1(x1)$$

$$y1'(4.8) = 16.051$$

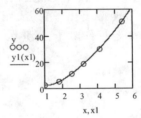

Example set 6.1

1. In a transient mass transfer of CO_2 through a membrane, the concentration (C in $kgmol/m^3$) profile of CO_2 at a certain time t across the membrane wall can be represented by the following equation:

$$C(x) = -1.3 \cdot 10^6 \cdot x^4 + 7.1 \cdot 10^3 \cdot x^3 - 14.0 x^2 - 0.364 x + 0.001 \qquad -0.001 \le x \le 0.001$$

where x is the distance in m from the center of the membrane wall.

The diffusivity of CO_2 through the membrane is $3.26 \ 10^{-8} \ m^2/s$ and can be assumed constant. Is the amount of CO_2 in the membrane accumulating or depleting?

At time t, the accumulation of CO_2 takes place if the input mass transfer flux is larger than the output flux:

$$N_{CO2} \ (\text{at } x = -0.001) - N_{CO2} \ (\text{at } x = 0.001) > 0$$

and the depletion of CO_2 takes place if the input flux is less than the output flux:

$$N_{CO2} \ (\text{at } x = -0.001) - N_{CO2} \ (\text{at } x = 0.001) < 0$$

The mass transfer flux (N_{CO2} in $kgmol/(s.m^2)$) at x m from the center of the membrane is given by

$$N_{CO2} = -D \cdot \frac{dC(x)}{dx}$$

Solution:

$$D_{CO2} := 3.26 \, 10^{-8} \quad m^2/s$$

$$C(x) := -1.3 \cdot 10^6 \cdot x^4 + 7.1 \cdot 10^3 \cdot x^3 - 14.0 x^2 - 0.364 x + 0.001$$

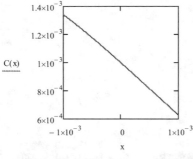

$$N_{CO2}(x) := -D_{CO2} \frac{d}{dx} C(x)$$

$$N_{CO2}(-0.001) = 1.009 \times 10^{-8} \quad kgmol/(m^2.s)$$

$$N_{CO2}(0.001) = 1.225 \times 10^{-8} \quad kgmol/(m^2.s)$$

$$N_{CO2}(-0.001) - N_{CO2}(0.001) = -2.165 \times 10^{-9} \quad kgmol/(m^2.s) \qquad depletion$$

2. Data for filtration of $CaCO_3$ slurry in water at 298 K at a constant pressure ($-\Delta p$) of 3 bar are as follows:

t [sec]	4.4	9.5	16.3	24.6	34.7	46.1	59.0	73.6	89.4	107.3
V [liter]	0.498	1.000	1.501	2.000	2.498	3.002	3.506	4.004	4.502	5.009

The filter area of the plate-and-frame press (A) is 400 cm² and the slurry concentration (c_s) is 20.0 g/liter. Estimate the specific cake resistance (α) in m/kg and the resistance of the filter medium to filtrate flow (R_m) in m⁻¹. At 298 K, the viscosity of water is 8.937 10⁻⁴ Pa.s.

The equations for constant-pressure filtration are as follows:

$$\frac{dt}{dV} = \frac{\mu \cdot \alpha \cdot c_s}{A^2 \cdot (-\Delta p)} \cdot V + \frac{\mu \cdot Rm}{A \cdot (-\Delta p)}$$

where μ is the viscosity of filtrate (water).

Solution:

$$vt := \begin{pmatrix} 4.4 \\ 9.5 \\ 16.3 \\ 24.6 \\ 34.7 \\ 46.1 \\ 59.0 \\ 73.6 \\ 89.4 \\ 107.3 \end{pmatrix} \cdot s \quad vV := \begin{pmatrix} 0.498 \\ 1.000 \\ 1.501 \\ 2.000 \\ 2.498 \\ 3.002 \\ 3.506 \\ 4.004 \\ 4.502 \\ 5.009 \end{pmatrix} \cdot liter$$

$$vs := cspline(vV, vt)$$

$$y(x) := interp(vs, vV, vt, x)$$

$$x := 0 \cdot liter, 0.01 \cdot liter .. 6 \cdot liter$$

The derivative of t with respect to V is a linear function of V.

$$f(x) := \frac{d}{dx} y(x) \qquad a_1 := slope\left(vV, \overrightarrow{f(vV)}\right)$$

$$a_0 := intercept\left(vV, \overrightarrow{f(vV)}\right)$$

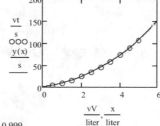

$$a_0 = 5.46 \times 10^3 \frac{s}{m^3} \qquad a_1 = 6.312 \times 10^6 \frac{s}{m^6}$$

$$g(x) := a_0 + a_1 \cdot x \qquad r := corr\left(\overrightarrow{f(vV)}, \overrightarrow{g(vV)}\right) = 0.999$$

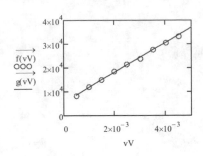

$$\mu := 8.937 \cdot 10^{-4} \cdot Pa \cdot s \qquad minus\Delta p := 3 \cdot bar$$

$$c_s := 20 \cdot \frac{gm}{liter} \qquad A := 400 \cdot cm^2$$

$$Rm := \frac{a_0 \cdot A \cdot minus\Delta p}{\mu} = 7.331 \times 10^{10} \frac{1}{m}$$

$$\alpha := \frac{a_1 \cdot A^2 \cdot minus\Delta p}{\mu \cdot c_s} = 1.695 \times 10^{11} \frac{m}{kg}$$

6.2 Integration

Procedure 6.4 shows the symbolic (analytical) integration of known function, Procedure 6.5 shows the numerical integration of known function using an integral operator, and Procedure 6.6 shows the integration from a set of data.

Procedure 6.4: *Symbolic Integration*

For example we need to calculate the integral of the following function from x = 0.1 to 0.5:

$$f(x) := 3 \cdot x \cdot exp(2 \cdot x)$$

1. Use the integral operator from the Calculus Toolbox and the symbolic equal sign and assign the result to a function, for example Int(a,b):

$$Int(a,b) := \int_a^b f(x)\, dx \rightarrow \frac{3 \cdot e^{2 \cdot a}}{4} - \frac{3 \cdot e^{2 \cdot b}}{4} - \frac{3 \cdot a \cdot e^{2 \cdot a}}{2} + \frac{3 \cdot b \cdot e^{2 \cdot b}}{2}$$

2. To calculate the integral of f(x) from x = 0.1 to 0.5, we can just use the function above:

$$Int(0.1, 0.5) = 0.733$$

Procedure 6.5: *Numerical Integration*

For example we need to calculate the integral of the following function from x = 0.1 to 0.5 and y = 0.2 to 0.4:

$$f(x,y) := \frac{x^3 \cdot y + 3 \cdot x + sin(x) + 3 \cdot \sqrt{x \cdot y} \cdot exp\left(2 \cdot x^2\right)}{\left(y + x^2 \cdot tan(x)\right) \cdot ln(x) - x \cdot y}$$

1. Use the integral operator from the Calculus Toolbox and directly assign to a new function:

$$Int(a,b,c,d) := \int_c^d \int_a^b f(x,y)\, dx\, dy$$

2. Calculate the integral from x = 0.1 to 0.5 and y = 0.2 to 0.4:

$$Int(0.1, 0.5, 0.2, 0.4) = -0.391$$

Procedure 6.6: *Integration of Data*

For example we need to determine the integral of y from x = 2 to 4 using the following data:

$$x := \begin{pmatrix} 1 \\ 1.8 \\ 2.5 \\ 3.2 \\ 4.1 \\ 5.4 \end{pmatrix} \qquad y := \begin{pmatrix} 2.5 \\ 8.1 \\ 10.0 \\ 18.5 \\ 33.0 \\ 50.7 \end{pmatrix}$$

1. Plot y vs. x to observe the smoothness of the data. From the plot below, we know that the data are not smooth. Thus, we could use polynomial regression to fit the data.

$k := 2$

$v := \text{regress}(x, y, k)$

$yl(xl) := \text{interp}(v, x, y, xl)$

$r := \text{corr}\left(y, \overrightarrow{yl(x)}\right)$ $r = 0.996$

2. Plot the fitting curve to check the goodness of fit visually

3. Calculate the integral:

$$Int(a, b) := \int_a^b yl(xl)\,dxl$$

$Int(2, 4) = 35.272$

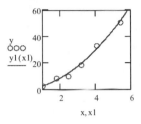

Example set 6.2

1. The velocity profile of an incompressible laminar flow through a circular pipe of radius R is given by:

$$v(r) = v_{max}\left[1 - \left(\frac{r}{R}\right)^2\right]$$ r = the distance from the center of the pipe

Determine the velocity at the center (v_{max}) of this flow if the diameter of the pipe is 2 in and the average velocity (v_{ave}) is 1 cm/sec. Then plot the velocity profile in the pipe (as a function of r). The velocity at the center is related to the average velocity by the following equation:

$$v_{max} = \frac{v_{ave} \cdot \pi \cdot R^2}{\int_0^{2\cdot\pi}\int_0^R \left[1 - \left(\frac{r}{R}\right)^2\right] \cdot r\,dr\,d\theta}$$

Solution: $R := 2 \cdot in$ $v_{ave} := 1 \cdot \dfrac{cm}{sec}$ $v_{max} := \dfrac{v_{ave} \cdot \pi \cdot R^2}{\int_0^{2\cdot\pi}\int_0^R \left[1 - \left(\frac{r}{R}\right)^2\right] \cdot r\,dr\,d\theta}$

$v_{max} = 2\,\dfrac{cm}{sec}$

$v(r) := v_{max}\left[1 - \left(\dfrac{r}{R}\right)^2\right]$ $r := 0 \cdot in, 0.01 \cdot in \ .. \ 2 \cdot in$

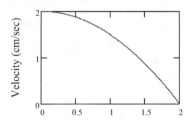

Distance from the center (in)

2. The heat capacity (C_p) of a gas is tabulated as a function of temperature (t):

t (°C)	20	50	80	110	140	170	200	230
Cp (J/mol°C)	28.95	29.13	29.30	29.48	29.65	29.82	29.99	30.16

Calculate the enthalpy change (ΔH) for 5 mol (n) of this gas going from 60 (t_1) to 180°C (t_2).

The enthalphy change is calculated from: $\Delta H = n \cdot \displaystyle\int_{t_1}^{t_2} C_p\,dt$

Solution:

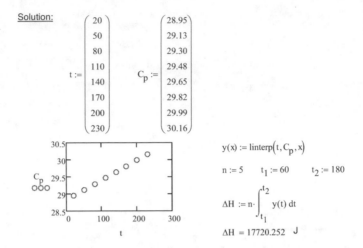

$$t := \begin{pmatrix} 20 \\ 50 \\ 80 \\ 110 \\ 140 \\ 170 \\ 200 \\ 230 \end{pmatrix} \qquad C_p := \begin{pmatrix} 28.95 \\ 29.13 \\ 29.30 \\ 29.48 \\ 29.65 \\ 29.82 \\ 29.99 \\ 30.16 \end{pmatrix}$$

$y(x) := \text{linterp}(t, C_p, x)$

$n := 5 \qquad t_1 := 60 \qquad t_2 := 180$

$$\Delta H := n \cdot \int_{t_1}^{t_2} y(t)\, dt$$

$\Delta H = 17720.252 \quad \text{J}$

Problems

1. A theoretical equation of state for hard-chain fluids, which is derived from thermodynamic perturbation theory, is given in term of the residual Helmholtz energy

    $$\tilde{a}^{res} = m \frac{4\eta - 3\eta^2}{(1-\eta)^2} + (1-m)\ln\frac{1-\eta/2}{(1-\eta)^3}$$

    where m is the number of segments in a chain and η is the reduced density given by

    $$\eta = m\frac{\pi}{6}N_A\rho\sigma^3$$

    In the equation above, N_A is the Avogadro number, ρ is the molar density, and σ is the hard-sphere segment diameter.

    The compressibility factor $Z\,(= Pv/RT)$ of any fluids can be calculated from the residual Helmholtz energy using the following equation:

    $$Z = 1 + \eta\frac{\partial\tilde{a}^{res}}{\partial\eta}$$

    a. Plot the compressibility factor as a function of the reduced density, say from 0 to 0.4, for a hard-chain fluid consisting 5 segments per chain.

    b. What is the compressibility factor of this fluid at $\rho = 0.003$ mol/cm³? Each segment has a diameter of 3.5 Å.

2. A mixture of 200 mol containing 60 mol% n-pentane and 40 mol% n-heptane is distilled in a tank under differential conditions at 1 bar until 120 mol has been distilled. What is the composition of the liquid

left in the tank?

The equilibrium data are as follows:

x	1.000	0.867	0.594	0.398	0.254	0.145	0.059	0.000
y	1.000	0.984	0.925	0.836	0.701	0.521	0.271	0.000

where x and y are mol fractions of n-pentane in the liquid and vapor phases, respectively.

The following equation is derived from a material balance for simple differential distillation:

$$\ln\left(\frac{L_1}{L_0}\right) = \int_{x_0}^{x_1} \frac{1}{y-x} dx$$

where L_0 is the number of moles of liquid originally charged, L_1 is the number of moles of liquid left in the tank, x_0 is the initial mole fraction of n-pentane, and x_1 is the final mole fraction of n-pentane.

3. Calculate the vapor pressure of carbon dioxide at 283 K using Peng-Robinson equation of state described in Example 1 of Example set 3.1. Graphically, vapor pressure is a pressure at which area I, shown in the Pressure-Volume (PV) plot below, is exactly equal to area II. P_{PR} is the pressure calculated from Peng-Robinson equation of state

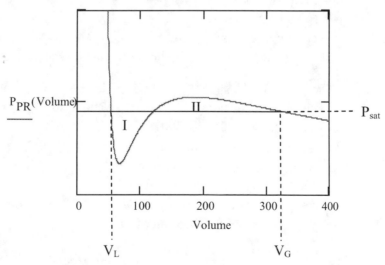

Mathematically, vapor pressure is a pressure (P) at which the following equation is satisfied:

$$P(V_G - V_L) = \int_{V_L}^{V_G} P_{PR} dV$$

where V is the molar volume and the subscripts G and L denote the gas and liquid phases, respectively.

Hint. Since calculating molar volumes of gas and liquid requires pressure, you need to create functions to describe the molar volumes of gas and liquid phases (as functions of pressure).

4. In an experiment, the following data were obtained for a gas phase reaction:

t (min)	0.0	2.5	5.0	10.0	15.0	20.0
P (mmHg)	7.8	10.5	12.7	15.8	17.9	19.4

This reaction can be modeled by using the following equation:

$$\frac{dP}{dt} = k(3P_0 - P)^\alpha$$

where t is time, P is the total pressure, P_0 is the initial pressure, k is the reaction constant and α is the order of reaction. From the data obtained, determine the reaction constant and the order of reaction. What is the unit of k?

5. A packed countercurrent cooling tower operating at 1 atm (P) will be designed to cool the existing process water. The flow rate and temperatures of the entering and leaving water have been dictated by the process requirement. If the dry air flow rate (G) is designed to be 1.356 kg/(s.m²), calculate the height of the tower (z).

At the condition of interest, the following data are known:

1. The volumetric mass-transfer coefficient in the gas phase (k_G) is 1.207×10^{-7} kgmol/(s.m³.Pa)
2. The enthalpy of the entering air (H_1) is 71.7×10^3 J/kg
3. The enthalpy of the leaving air (H_2), which is obtained from the heat balance, is 129.9×10^3 J/kg
4. The enthalpy of air at the gas interface (H_i) as a function of the enthalpy of air (H) is tabulated as follows:

H [10^3 J/kg]	71.7	83.5	94.9	106.5	118.4	129.9
H_i [10^3 J/kg]	94.4	108.4	124.4	141.8	162.1	184.7

The height of the tower can be calculated from:

$$z = \frac{G}{M_{air} P k_G} \int_{H_1}^{H_2} \frac{1}{H_i - H} dH$$

where M_{air} is the molecular weight of air [= 29.0 kg/kgmol].

6. In a material fabrication process, an oxidized-nickel cylinder 3 cm in diameter and 10 cm in length is cooled in a vacuum chamber, the wall temperature of which is kept constant at 300 K (T_w). The initial temperature of this cylinder is 1000 K (T_i) and the target final temperature is 600 K (T_f). Estimate the time needed for the cooling process.

The properties of nickel at the average temperature of the process are density (ρ) = 8900 kg/m³, specific heat (c_p) = 530 J/(kg·K), and emissivity (ε) = 0.49.

For this condition, the time needed can be calculated from:

$$t = -\frac{\rho V c_p}{\varepsilon \sigma A} \int_{T_i}^{T_f} \frac{dT}{T^4 - T_w^4}$$

where σ is the Boltzmann constant [= 5.67×10^{-8} W/(m²K⁴)], V is the volume of the cylinder, and A is the surface area of the cylinder.

7. In a certain application involving parallel airflow over a flat plate of length 2 m (L), the spatial variation of temperatures measured in the boundary layer is accurately correlated by the following expression:

$$T = 30 + 47 \exp\left(-500 \frac{y}{\sqrt{x}}\right) \quad [°C]$$

where x is the distance parallel to the plate measured from the leading edge [in m] and y is the distance normal to the plate measured from the plate surface [in m].

a. Estimate the free stream air temperature (T_∞), i.e., at large y, and the surface temperature (T_s).

b. What is the local heat flux at $x = 1$ m (assume thermal conductivity $k = 0.03$ W/(m·K))?

c. What is the average heat flux from this surface?

The local heat flux is given by:

$$q_x'' = -k \frac{\partial T}{\partial y}\bigg|_{y=0}$$

The average heat flux is given by:

$$q_{av}'' = \overline{h}(T_s - T_\infty)$$

where $\overline{h}$ is the average convection heat transfer coefficient obtained from

$$\bar{h} = \frac{1}{L}\int_0^L h_x\, dx \quad \text{and} \quad h_x = \frac{-k\left.\dfrac{\partial T}{\partial y}\right|_{y=0}}{T_s - T_\infty}$$

Hint: For this fast varying function, i.e., T changes greatly with y, numerical derivative operator is not accurate. Thus, use symbolic differentiation.

8. In the development of an artificial kidney that could be worn by the patient, a reaction that converts urea to ammonia and carbon dioxide by using enzyme urease needs to be studied. One set of experimental data using a certain amount of urease is as follows:

C, kmol/m³	0.2	0.02	0.01	0.005	0.002
$-r_{urea}$, mol/(m³·s)	1.08	0.55	0.38	0.2	0.09

In the table above, C is the concentration of urea and $-r_{urea}$ is the rate of urea conversion. : The time needed to reduce the concentration of urea from C_0 to C_f is given by:

$$t = \int_{C_f}^{C_0} \frac{dC}{-r_{urea}}$$

From reaction kinetics, it is known that the following model, referred to as Michaelis-Menten equation, relates the rate of urea conversion to the concentration of urea:

$$-r_{urea} = \frac{V_m C}{K_m + C}$$

where V_m and K_m are constants for a certain amount of urease.

Estimate the time needed to reduce the concentration of urea from 0.5 to 0.2 kmol/m³ using the same amount of urease as that in the above experiment. What are the units of V_m and K_m?

9. A gaseous stream consisting of 40 mol% ammonia ($y_1 = 0.4$) in an inert gas is scrubbed with water in a packed column. The inert gas molar flow rate (V) is 900 lbmol/h. The diameter of the column is 4 ft. Under the operating conditions, the estimated value of the gas film mass transfer coefficient ($k_y'a$) is assumed constant at 30 lbmol/(h·ft³). The desired ammonia concentration in the outlet gas stream is 3.2 mol% ($y_2 = 0.032$).

a. Determine the height of packed bed needed (z), which can be calculated from:

$$z = \frac{V'}{k_y' aS} \int_{y_2}^{y_1} \frac{(1-y)_{iM}}{(y-y_i)(1-y)^2} dy$$

with $(1-y)_{iM} = \dfrac{(1-y)-(1-y_i)}{\ln\left(\dfrac{1-y}{1-y_i}\right)}$

where S is the cross sectional area of the column and y_i is the mol fraction of ammonia at the water-gas interface, which in turn depends on the mol fraction of ammonia in the bulk gas phase (y) as shown in the data listed below:

y	0.032	0.080	0.125	0.166	0.206	0.243
y_i	0.004	0.029	0.056	0.084	0.115	0.148

y	0.278	0.311	0.342	0.372	0.400
y_i	0.183	0.221	0.261	0.304	0.349

b. Plot the height of packed bed as a function of the desired ammonia concentration in the outlet gas stream ($0.032 \le y_2 \le 0.20$).

10. In the design of a tray tower for absorption, for a certain solute-free liquid molar flow rate (L), the operating curve of the absorption operation, which relates the solute mole fractions in the bulk gas and liquid phases, is given by:

$$y = \frac{\dfrac{L'}{V'}\dfrac{x}{1-x} + \left(\dfrac{y_t}{1-y_t} - \dfrac{L'}{V'}\dfrac{x_t}{1-x_t}\right)}{1 + \dfrac{L'}{V'}\dfrac{x}{1-x} + \left(\dfrac{y_t}{1-y_t} - \dfrac{L'}{V'}\dfrac{x_t}{1-x_t}\right)}$$

where y is the mole fraction of solute in the bulk gas phase, x is the mole fraction of solute in the bulk liquid phase, L' is the molar flow rate of solute-free liquid, V' is the molar flow rate of solute-free gas ($= 85$ kmol/hr), and y_t ($= 0.01$) and x_t ($= 0.002$) are the mole fractions of solute in the bulk gas and liquid phases, respectively, at the top of the column. This operating curve is used in conjunction with the equilibrium curve.

Equilibrium data for the system of interest are given below:

x	0	0.033	0.072	0.117	0.171
y^*	0	0.0396	0.0829	0.1127	0.136

a. From the given equilibrium data above, use a curve fitting method to fit the data. Plot the data and the fitting curve ($0 \leq x \leq$
   0.18).

b. In the same graph, plot three operating curves for $L' = 170, 150$,
   and 130 kmol/hr.

c. In the design of an absorption tray tower, it is crucial to determine the minimum solute-free liquid flow rate that can be used.
   Any operations using liquid flow rate close to this minimum liquid flow rate require many trays. When minimum liquid flow
   rate is used, the operating curve becomes tangent to the equilibrium curve (the fitting curve obtained in part a). In other words,
   the operating and equilibrium curves have a common tangent line
   at a point. Determine this minimum solute-free liquid flow rate.
   Hint: To solve this problem, use the criteria of tangency of two
   curves.

d. To check whether the calculated minimum solute-free liquid flow
   rate is correct or not, plot the equilibrium curve along with the
   operating curve when the minimum liquid flow rate obtained
   from part c is used.

Chapter 7
Optimization

The important task of chemical engineer is to design new, better, more efficient, more environmental-benign, and more economical process systems as well as to improve the operation of existing process systems. This important task calls for a deeper understanding of chemical engineering principles, which underlie all physical and chemical phenomena occurring in processes and operations, and optimization methods to determine the best possible solution.

There are several types of optimization problems we might encounter in engineering, for example the extreme value problem, linear programming, and non-linear programming, which will be discussed in this chapter.

7.1. Extreme Value Problem

An extreme value problem is an optimization, either minimization or maximization, of an objective function without any constraints or with constraints that can be solved using stationary point approach.

Although Mathcad has very versatile functions for solving optimization problems, as shown later, we still discuss conventional approaches using stationary point. These conventional approaches give us better understanding of the issues arose in optimization problems.

7.1.1. Unconstrained Function

For optimization of a function without any constraints, the solutions are conventionally obtained by finding the stationary points, i.e., equalizing the first derivatives of the function with respect to the independent variables to zero and solving these equations, the procedure of which has been described in Chapters 3 and 4.

In order to determine the types of stationary points, i.e., maximum, minimum, or saddle point, we need to construct a matrix called Hessian matrix at the stationary point. For optimizing a function of n independent variables, the Hessian matrix is an $n \times n$ matrix:

$$\mathbf{H} = \left[h_{ij} \right]_{n \times n}$$

the element of which is defined as follows

$$h_{ij} = \frac{\partial^2 f}{\partial x_i \partial x_j} \quad \text{evaluated at the stationary point}$$

where x_i and x_j are the independent variables. Then, the type of the stationary point is

> *minimum* if all of its eigenvalues are real and positive numbers (the Hessian matrix is positive definite)
>
> *maximum* if all of its eigenvalues are real and negative numbers (the Hessian matrix is negative definite)
>
> *saddlepoint* otherwise (the Hessian matrix is indefinite)

The eigenvalues of a matrix can be calculated easily using *eigenvals* function. Procedure 7.1 shows this stationary point approach.

As we can notice from Procedure 7.1, the solution of an extreme value problem using stationary point approach is quite tedious. In fact, Mathcad offers easy ways to solve optimization problems. We can use *minimize* function to find a minimum stationary point and *maximize* function to find a maximum stationary point. However, we cannot directly find saddle stationary point. Procedure 7.2 shows how to use these two functions. *Minimize/maximize* function supports unit. If the independent variables have different types of unit, the solution must be assigned to a column matrix.

If *minimize* or *maximize* function gives an error message: "Could not find a maximum/minimum", there are two possibilities: the initial guesses are not good, thus need to be changed, or the function does not have a maximum or a minimum.

Procedure 7.1: *Stationary point method for an unconstrained function*

Suppose we want to determine the values of the following function at the stationary points:

$$f(x, y) := \left(x^2 + y - 11\right)^2 + \left(x + y^2 - 7\right)^2$$

1. Whenever possible, plot the function to get the idea of the solutions

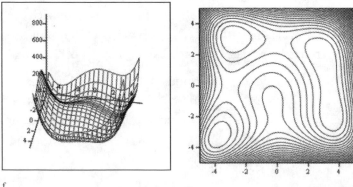

f   f

Surface and contour plots can be accessed from the Graph Toolbar. The function name we want to plot is typed in the empty place holder. The contour plot helps us to determine initial guesses.

2. Define functions for calculating all of the second derivatives needed in the Hessian matrix:

$$f_{xx}(x,y) := \frac{d^2}{dx^2}f(x,y) \quad f_{xy}(x,y) := \frac{d}{dx}\frac{d}{dy}f(x,y) \quad f_{yy}(x,y) := \frac{d^2}{dy^2}f(x,y)$$

3. Construct the Hessian matrix

$$H(x,y) := \begin{pmatrix} f_{xx}(x,y) & f_{xy}(x,y) \\ f_{xy}(x,y) & f_{yy}(x,y) \end{pmatrix}$$

4. Create a function for calculating the eigenvalues of the Hessian matrix:

$$Q(s) := \text{eigenvals}\left(H\left(s_0, s_1\right)\right)$$

5. Create a function to find stationary points using *Given* block and *find* function:

Given

$$\frac{d}{dx}f(x,y) = 0 \qquad \frac{d}{dy}f(x,y) = 0$$

$$s(x,y) := \text{Find}(x,y)$$

If the function is very complex, *find* function may not work. In this case try to replace the *find* function with *minerr* function, which will try to find the closest solution. *Minerr* function has the same arguments as *find* function.

Note that the arguments x and y on the left side are in fact the initial guesses, which are still left open. We want to use different initial guesses to obtain different roots.

6. Determine the stationary points, type of each stationary point, and the values of the function:

$$p := s(4,2) = \begin{pmatrix} 3 \\ 2 \end{pmatrix} \qquad Q(p) = \begin{pmatrix} 82.284 \\ 25.716 \end{pmatrix} \qquad f\left(p_0, p_1\right) = 0 \qquad \text{min}$$

$$p := s(3,0) = \begin{pmatrix} 3.385 \\ 0.074 \end{pmatrix} \qquad Q(p) = \begin{pmatrix} 97.548 \\ -14.135 \end{pmatrix} \qquad f\left(p_0, p_1\right) = 13.312 \qquad \text{saddle}$$

$$p := s(4,-2) = \begin{pmatrix} 3.584 \\ -1.848 \end{pmatrix} \qquad Q(p) = \begin{pmatrix} 105.419 \\ 28.691 \end{pmatrix} \qquad f\left(p_0, p_1\right) = 0 \qquad \text{min}$$

$$p := s(0,3) = \begin{pmatrix} 0.087 \\ 2.884 \end{pmatrix} \qquad Q(p) = \begin{pmatrix} -31.707 \\ 75.508 \end{pmatrix} \qquad f\left(p_0, p_1\right) = 67.719 \qquad \text{saddle}$$

$$p := s(0,-1) = \begin{pmatrix} -0.271 \\ -0.923 \end{pmatrix} \qquad Q(p) = \begin{pmatrix} -45.605 \\ -16.066 \end{pmatrix} \qquad f\left(p_0, p_1\right) = 181.617 \qquad \text{max}$$

$$p := s(-3,3) = \begin{pmatrix} -2.805 \\ 3.131 \end{pmatrix} \qquad Q(p) = \begin{pmatrix} 64.84 \\ 80.55 \end{pmatrix} \qquad f\left(p_0, p_1\right) = 0 \qquad \text{min}$$

$$p := s(-3,0) = \begin{pmatrix} -3.073 \\ -0.081 \end{pmatrix} \qquad Q(p) = \begin{pmatrix} 72.435 \\ -39.651 \end{pmatrix} \qquad f\left(p_0, p_1\right) = 104.015 \qquad \text{saddle}$$

$$p := s(-4,-4) = \begin{pmatrix} -3.779 \\ -3.283 \end{pmatrix} \qquad Q(p) = \begin{pmatrix} 133.786 \\ 70.714 \end{pmatrix} \qquad f\left(p_0, p_1\right) = 0 \qquad \text{min}$$

Procedure 7.2: *minimize/maximize for an unconstrained function*

Suppose we want to determine the values of the following function at the stationary points:

$$f(x,y) := \left(x^2 + y - 11\right)^2 + \left(x + y^2 - 7\right)^2$$

1. Whenever possible, plot the function to get the ideas of the solutions.

2. Create functions for calculating minimum and maximum stationary point using *minimize* and *maximize* functions:

$$\text{smin}(xi, yi) := \text{Minimize}(f, xi, yi)$$

$$\text{smax}(xi, yi) := \text{Maximize}(f, xi, yi)$$

xi and yi are the needed initial guesses, which are conveniently used as parameters.

3. Calculate the stationary points and the values of the function, for example:

$$\begin{pmatrix} x \\ y \end{pmatrix} := \text{smin}(4, 1) = \begin{pmatrix} 3 \\ 2 \end{pmatrix} \qquad f(x, y) = 8.948 \times 10^{-14}$$

$$\begin{pmatrix} x \\ y \end{pmatrix} := \text{smax}(0, 0) = \begin{pmatrix} -0.271 \\ -0.923 \end{pmatrix} \qquad f(x, y) = 181.617$$

If we would not change the initial guesses, we may not need to create the function in step 2, as shown below:

Initial guesses: $xi := 4$ $yi := 1$ $\begin{pmatrix} x \\ y \end{pmatrix} := \text{Minimize}(f, xi, yi) = \begin{pmatrix} 3 \\ 2 \end{pmatrix}$

7.1.2. Constrained Function

For optimization of a function subject to equality constraints, the solutions are conventionally obtained using the method of Lagrange Multipliers. Consider the minimization of a function of n variables, referred to as the objective function, subject to m equality constraints:

Minimize $f(x_1, x_2, ..., x_n)$

Subject to $\begin{cases} g_1(x_1, x_2, ..., x_n) = 0 \\ g_2(x_1, x_2, ..., x_n) = 0 \\ \qquad . \\ \qquad . \\ g_m(x_1, x_2, ..., x_n) = 0 \end{cases}$

The method of Lagrange Multipliers converts this problem to the following unconstrained optimization problem:

Minimize $L(\mathbf{x}, \lambda) = f(\mathbf{x}) - \sum_{i=1}^{m} \lambda_i g_i(\mathbf{x})$

which now has $n+m$ independent variables, i.e., x_1, x_2, ..., x_n, and λ_1, λ_2, ..., λ_m. λ's are called the multipliers.

The rest of the procedure is to find the stationary point(s), determine the type of the stationary point(s), and determine the values of the function at the stationary point(s). To test whether the stationary point corresponds to a minimum, a maximum, or a saddle point, we again use the Hessian matrix described in the optimization of unconstrained function. The Hessian matrix does not include the derivatives of the function with respect to the multipliers.

Procedure 7.3 shows how to optimize constrained function using the method of Lagrange Multipliers. In fact, the example in Procedure 7.3 can be transformed into an unconstrained optimization problem by solving for y (in term of x) from the constraint and then substituting to the objective function. Thus, the objective function will have only one independent variable, i.e., x. However, this cannot always be done, although it is always a good idea to try to reduce the number of the independent variables by substitution.

A simpler method is again to use m*inimize* and *maximize* functions. Procedure 7.4 demonstrates the use of *minimize /maximize* function for this purpose.

Procedure 7.3: *Lagrange Multipliers*

Suppose we want to solve the following optimization problem:

Minimize $\qquad$ $f(x,y) = (x-2)^2 + y^2$ $\qquad$ Subject to $\qquad$ $2 \cdot x + y - 2 = 0$

1. Create the function to be optimized and a function representing the constraint:

$$f(x,y) := (x-2)^2 + y^2 \qquad\qquad g(x,y) := 2 \cdot x + y - 2$$

2. Whenever possible, plot the objective function and constraint to get the idea of the solution:

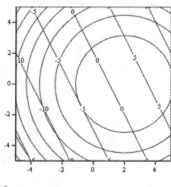

The numbered curves are the function g(x,y).

f, g

3. Construct the unconstrained function, i.e., the Lagrangian function L:

$$L(x,y,\lambda) := f(x,y) - \lambda \cdot g(x,y)$$

4. Define functions for calculating all of the second derivatives needed in the Hessian matrix:

$$L_{xx}(x,y,\lambda) := \frac{d^2}{dx^2}L(x,y,\lambda) \quad L_{xy}(x,y,\lambda) := \frac{d}{dx}\frac{d}{dy}L(x,y,\lambda) \quad L_{yy}(x,y,\lambda) := \frac{d^2}{dy^2}L(x,y,\lambda)$$

5. Construct the Hessian matrix $\qquad$ $H(x,y,\lambda) := \begin{pmatrix} L_{xx}(x,y,\lambda) & L_{xy}(x,y,\lambda) \\ L_{xy}(x,y,\lambda) & L_{yy}(x,y,\lambda) \end{pmatrix}$

6. Create a function for calculating the eigenvalues of the Hessian matrix:

$$Q(s) := eigenvals\left(H\left(s_0, s_1, s_2\right)\right)$$

7. Find a stationary point using *Given* block and *find* function:

Initial guesses: $\qquad$ $x := 0 \qquad y := 0 \qquad \lambda := 1$

Given

$$\frac{d}{dx}L(x,y,\lambda) = 0 \qquad \frac{d}{dy}L(x,y,\lambda) = 0 \qquad \frac{d}{d\lambda}L(x,y,\lambda) = 0$$

$$p := Find(x,y,\lambda) = \begin{pmatrix} 1.2 \\ -0.4 \\ -0.8 \end{pmatrix}$$

8. Determine the type of the stationary point and the value of the function:

$$Q(p) = \begin{pmatrix} 2 \\ 2 \end{pmatrix} \qquad minimum \qquad f\left(p_0, p_1\right) = 0.8 \qquad \text{Check the constraint:}$$

$$g\left(p_0, p_1\right) = 1.332 \times 10^{-15}$$

Procedure 7.4: _minimize/maximize for a constrained function_

Suppose we want to solve the following optimization problem:

Minimize $f(x,y) = (x-2)^2 + y^2$

Subject to $2 \cdot x + y - 2 = 0$

1. Create the function to be optimized: $f(x,y) := (x-2)^2 + y^2$

2. Whenever possible, plot the function to get the ideas of the solutions.

3. Construct a _Given_ block for the constraint and provide initial guesses:

Initial guesses: $x := 4$ $y := 2$

Given
$$2 \cdot x + y - 2 = 0$$

4. Calculate the minimum stationary point using _minimize_ function:

$$\begin{pmatrix} x \\ y \end{pmatrix} := \text{Minimize}(f,x,y) = \begin{pmatrix} 1.2 \\ -0.4 \end{pmatrix} \qquad f(x,y) = 0.8$$

Example set 7.1

1. Substance D is produced by decomposition reaction of A in a mixed reactor. The reaction is accompanied by two other undesired side reactions:

     B $r_B = 1$

A → C $r_C = 0.2 \cdot C_A$

     D (desired product) $r_D = 0.5 \cdot C_A^2$

C_A is the concentration of A in the exit stream, r_B, r_C, and r_D are the reaction rates of the production of B, C, and D, respectively.

From the kinetic considerations, the concentration of D in the exit stream is given by

$$C_D = \frac{r_D}{r_B + r_C + r_D} \cdot (C_{A0} - C_A) \qquad C_{A0} = 3 \quad \text{mol/liter}$$

Determine the optimum concentration of A in the exit stream that maximizes the desired product D. Also, what is the maximum concentration of D at this condition?

<u>Solution:</u> $C_{A0} := 3$ $C_D(C_A) := \dfrac{0.5 \cdot C_A^2}{1 + 0.2 \cdot C_A + 0.5 \cdot C_A^2} \cdot (C_{A0} - C_A)$

Initial guess: $C_A := 1$

$C_A := \text{Maximize}(C_D, C_A)$

$C_A = 1.473$ mol/liter   (optimum C_A)

$C_D(C_A) = 0.696$ mol/liter   (maximum product)

If conventional method is used (by using derivative):

Initial guess: $C_A := 1$ Given

$$\frac{d}{dC_A} C_D(C_A) = 0$$

$$C_A := \text{Find}(C_A) \qquad C_A = 1.473 \qquad C_D(C_A) = 0.696$$

Test the stationary point: $\dfrac{d^2}{dC_A^2} C_D(C_A) = -0.889$ maximum

2. Synthesis gas can be produced by catalytic reforming of methane (CH_4) with steam at high temperature and atmospheric pressure:

$$CH_4 \ (1) + 2H_2O \ (2) \ \text{-------->} \ CO_2 \ (4) + 4H_2 \ (5)$$

The water-gas-shift reaction also occurs along with the above reaction:

$$CO_2 + H_2 \ \text{-------->} \ CO \ (3) + H_2O$$

If the catalytic reforming is carried out at 1000 K and 1 bar, determine the equilibrium compositions (mol fractions) of the mixture. The reactants are supplied to the reactor in the ratio of 2 mol steam to 1 mol methane.

The equilibrium compositions can be calculated by minimizing the total Gibbs free energy of the system (G_t):

Minimize $\quad G_t\left(n_1, n_2, n_3, n_4, n_5\right)$

Subject to material balances:
$$n_1 + n_3 + n_4 = 1$$
$$4 \cdot n_1 + 2 \cdot n_2 + 2 \cdot n_5 = 8$$
$$n_2 + n_3 + 2 \cdot n_4 = 2$$

n_i is the number of moles of component i in the mixture

The mol fraction of component i and the total Gibbs free energy of the systemis can be calculated from:

$$y_i = \frac{n_i}{\sum_j n_j} \qquad G_t = \sum_i \left(n_i \cdot \Delta G_i\right) + R \cdot T \cdot \sum_i \left(n_i \cdot \ln\left(\frac{n_i}{\sum_j n_j}\right)\right)$$

where the summation is over all components in the system, R is the gas constant, T is the absolute temperature, and ΔG_i is the standard Gibbs energy change of formation for component i. The standard Gibbs energy change of formation at 1000 K for each of the component is as follows:

$$\Delta G_1 = 19720 \, J \cdot mol^{-1} \qquad \Delta G_2 = -192420 \, J \cdot mol^{-1} \qquad \Delta G_3 = -200240 \, J \cdot mol^{-1}$$

$$\Delta G_4 = -395790 \, J \cdot mol^{-1} \qquad \Delta G_5 = 0 \cdot J \cdot mol^{-1}$$

Solution: $\quad$ CH4 (1) $\quad$ H2O (2) $\quad$ CO (3) $\quad$ CO2 (4) $\quad$ H2 (5) $\qquad$ ORIGIN:= 1

$$\Delta G := \begin{pmatrix} 19720 \\ -192420 \\ -200240 \\ -395790 \\ 0 \end{pmatrix} \qquad i := 1..5 \quad N(n) := \sum_i n_i \qquad R := 8.314472 \qquad T := 1000$$

$$G_t(n) := \sum_i \left(n_i \cdot \Delta G_i\right) + R \cdot T \cdot \sum_i \left(n_i \cdot \ln\left(\frac{n_i}{N(n)}\right)\right)$$

Initial guesses: $\quad n_i := 1$

Given

$$n_1 + n_3 + n_4 = 1 \qquad 4 \cdot n_1 + 2 \cdot n_2 + 2 \cdot n_5 = 8 \qquad n_2 + n_3 + 2 \cdot n_4 = 2$$

$n := \text{Minimize}\left(G_t, n\right)$

$$n = \begin{pmatrix} 0.041 \\ 0.785 \\ 0.702 \\ 0.257 \\ 3.133 \end{pmatrix} \qquad y_i := \frac{n_i}{N(n)} \qquad y = \begin{pmatrix} 8.379 \times 10^{-3} \\ 0.16 \\ 0.143 \\ 0.052 \\ 0.637 \end{pmatrix}$$

7.2 Linear Programming

Many operations management decisions attempt to make the most effective use of limited resources and obtain minimum operating cost or maximum production and profit. Resources typically include money, time, machinery, labor, and raw materials. Linear programming is a widely used technique designed to help engineers plan and make decisions necessary to allocate resources.

Linear Programming is an optimization of a linear objective function with linear inequality and/or equality constraints. Again, we can use *maximize/minimize* function to solve this type of problem. Procedure 7.5 describes how to use these functions, which is similar to that has been discussed in the previous section.

Procedure 7.5: _minimize/maximize for Linear Programming_

Suppose we want to solve the following optimization problem:

Maximize $f(x, y) = 3 \cdot x + 2 \cdot y$

Subject to $-x + 2 \cdot y \leq 4$ $0 \leq x \leq 3$ $x - y \leq 3$ $3 \cdot x + 2 \cdot y \leq 14$ $y \geq 0$

1. Define the function: $f(x, y) := 3 \cdot x + 2 \cdot y$

2. Give initial guesses: $x := 0$ $y := 0$

3. Set up the constraints in a *Given* block:

Given

$-x + 2 \cdot y \leq 4$ $0 \leq x \leq 3$ $x - y \leq 3$ $3 \cdot x + 2 \cdot y \leq 14$ $y \geq 0$

$\begin{pmatrix} x \\ y \end{pmatrix} := \text{Maximize}(f, x, y) = \begin{pmatrix} 3 \\ 2.5 \end{pmatrix}$ $f(x, y) = 14$

Example set 7.2

1. Five crude oils of different grades are processed in a refinery to produce four different products: gasoline, heating oil, jet fuel, and lube oil. The following table shows the fractions of products that can be obtained from a crude oil:

Crude Oil	Gasoline	Heating oil	Jet fuel	Lube oil
	bbl product per bbl of crude[‡]			
Crude 1	0.6	0.2	0.1	0
Crude 2	0.5	0.2	0.2	0
Crude 3	0.3	0.3	0.3	0
Crude 4	0.4	0.3	0.2	0
Crude 5	0.4	0.1	0.2	0.2

[‡]The fractions do not add up to 1 due to losses in processing

Other data:

Crude Oil	Cost, $/bbl	Operating cost, $/bbl	Availability, bbl/week
Crude 1	45.00	12.00	80,000
Crude 2	43.00	20.00	100,000
Crude 3	40.00	17.00	100,000
Crude 4	52.00	7.50	100,000
Crude 5	65.00	6.50	60,000

Product	Price, $/bbl	Maximum demand, bbl/week
Gasoline	105.00	170,000
Heating oil	95.00	85,000
Jet fuel	61.00	75,000
Lube oil	140.00	30,000

For maximum profit, determine the amount of each crude oil to be processed per week. What is the profit per week?

This is an optimization (linear programming) problem:

Maximize $\quad P\left(x_1, x_2, x_3, x_4, x_5\right) \quad$ where P is the profit/week and x_i is the crude

Subject to $\quad$ Availability: $\quad x_1 \leq 80000 \quad x_2 \leq 100000 \quad x_3 \leq 100000$

$$x_4 \leq 100000 \quad x_5 \leq 60000$$

Maximum product demand (what can be sold):

$$0.6 \cdot x_1 + 0.5 \cdot x_2 + 0.3 \cdot x_3 + 0.4 \cdot x_4 + 0.4 \cdot x_5 \leq 170000$$

$$0.2 \cdot x_1 + 0.2 \cdot x_2 + 0.3 \cdot x_3 + 0.3 \cdot x_4 + 0.1 \cdot x_5 \leq 85000$$

$$0.1 \cdot x_1 + 0.2 \cdot x_2 + 0.3 \cdot x_3 + 0.2 \cdot x_4 + 0.2 \cdot x_5 \leq 75000$$

$$0.2 \cdot x_5 \leq 30000$$

Positive variables $\quad x_1 \geq 0 \quad x_2 \geq 0 \quad x_3 \geq 0 \quad x_4 \geq 0 \quad x_5 \geq 0$

Solution: $\quad$ ORIGIN:= 1

Set up the objective function: (note that the equation can be typed in several lines using [Ctrl][Enter])

$$P(x) := \left(0.6 \cdot x_1 + 0.5 \cdot x_2 + 0.3 \cdot x_3 + 0.4 \cdot x_4 + 0.4 \cdot x_5\right) \cdot 105 \dots$$
$$+ \left(0.2 \cdot x_1 + 0.2 \cdot x_2 + 0.3 \cdot x_3 + 0.3 \cdot x_4 + 0.1 \cdot x_5\right) \cdot 95 \dots$$
$$+ \left(0.1 \cdot x_1 + 0.2 \cdot x_2 + 0.3 \cdot x_3 + 0.2 \cdot x_4 + 0.2 \cdot x_5\right) \cdot 61 + 0.2 \cdot x_5 \cdot 140 \dots$$
$$+ -\left(45 \cdot x_1 + 43 \cdot x_2 + 40 \cdot x_3 + 52 \cdot x_4 + 65 \cdot x_5\right) - \left(12 \cdot x_1 + 20 \cdot x_2 + 17.0 \cdot x_3 + 7.5 \cdot x_4 + 6.5 \cdot x_5\right)$$

Initial guesses: $\quad x := \begin{pmatrix} 80000 \\ 100000 \\ 100000 \\ 100000 \\ 60000 \end{pmatrix}$

Given

$$x_1 \leq 80000 \quad x_2 \leq 100000 \quad x_3 \leq 100000 \quad x_4 \leq 100000 \quad x_5 \leq 60000$$

$$0.6 \cdot x_1 + 0.5 \cdot x_2 + 0.3 \cdot x_3 + 0.4 \cdot x_4 + 0.4 \cdot x_5 \leq 170000$$

$$0.2 \cdot x_1 + 0.2 \cdot x_2 + 0.3 \cdot x_3 + 0.3 \cdot x_4 + 0.1 \cdot x_5 \leq 85000$$

$$0.1 \cdot x_1 + 0.2 \cdot x_2 + 0.3 \cdot x_3 + 0.2 \cdot x_4 + 0.2 \cdot x_5 \leq 75000$$

$$0.2 \cdot x_5 \leq 20000 \quad\quad x \geq 0$$

$$x := \text{Maximize}(P, x) \quad\quad x^T = (80000 \quad 90000 \quad 70000 \quad 80000 \quad 60000)$$

Crude 1: $x_1 = 80000$ bbl/week $\quad$ Crude 4: $x_4 = 80000$ bbl/week

Crude 2: $x_2 = 90000$ bbl/week $\quad$ Crude 5: $x_5 = 60000$ bbl/week

Crude 3: $x_3 = 70000$ bbl/week $\quad$ Profit: $\quad P(x) = 8910000 \quad$ $/week

2. A company manufactures three chemical products: product 1, product 2, and product 3. The needed resources of each product is shown in the following table:

Product	Engineering service (hr)	Direct labor (hr)	Raw material (lb)
1	1	10	3
2	2	4	2
3	1	5	1

There are 100 hr of engineering, 600 hr of labor, and 300 lb of material available per day. The unit profits on these products are $10, $6, and $4, respectively.

a. Determine the units of each product to be produced for maximum profit per day. What is the maximum profit?

b. What is the unit profit of product 3 that must be increased to before it becomes economical to produce?

c. If it is possible to increase the direct labor hours by 10% by scheduling overtime that requires an additional labor cost of $60/day, analyze whether this labor increase is economical or not.

This is a Linear Programming problem:

Maximize $f(x_1, x_2, x_3)$ (profit per day)

Subject to $x_1 + 2 \cdot x_2 + x_3 \le 100$ (engineering service)

                  $10 \cdot x_1 + 4 \cdot x_2 + 5 \cdot x_3 \le 600$ (direct labor)

                  $3 \cdot x_1 + 2 \cdot x_2 + x_3 \le 300$ (raw material)

                  $x_1 \ge 0 \quad x_2 \ge 0 \quad x_3 \ge 0$

Solution: ORIGIN:= 1

a. Objective function: $f(x) := 10 \cdot x_1 + 6 \cdot x_2 + 4 \cdot x_3$

Initial guesses: $x := \begin{pmatrix} 0 \\ 0 \\ 0 \end{pmatrix}$

Given

$\qquad x_1 + 2 \cdot x_2 + x_3 \le 100 \qquad\qquad x_1 \ge 0$

$\qquad 10 \cdot x_1 + 4 \cdot x_2 + 5 \cdot x_3 \le 600 \qquad x_2 \ge 0$

$\qquad 3 \cdot x_1 + 2 \cdot x_2 + x_3 \le 300 \qquad\qquad x_3 \ge 0$

$\qquad x := \text{Maximize}(f, x)$

$x = \begin{pmatrix} 50 \\ 25 \\ 0 \end{pmatrix}$ Product 1: $x_1 = 50$

                        Product 2: $x_2 = 25$ Profit: $PR1 := f(x)$

                        Product 3: $x_3 = 0$ $PR1 = 650$ $/day

b. As we can see from part (a), product 3 is not economical to produce when the unit profit is $4. To analyze this problem, it is better to write the objective function as a function of the unit profits too so that we can plot x_3 as a function of the unit profit of product 3:

$f(x, p1, p2, p3) := p1 \cdot x_1 + p2 \cdot x_2 + p3 \cdot x_3$

Given

Note that in the argument, the parameters must be typed after the independent variable x.

$\qquad x_1 + 2 \cdot x_2 + x_3 \le 100 \qquad\qquad x_1 \ge 0$

$\qquad 10 \cdot x_1 + 4 \cdot x_2 + 5 \cdot x_3 \le 600 \qquad x_2 \ge 0$

$\qquad 3 \cdot x_1 + 2 \cdot x_2 + x_3 \le 300 \qquad\qquad x_3 \ge 0$

The initial guesses are the x values obtained in part (a)

$x(p1, p2, p3) := \text{Maximize}(f, x)$

$p1 := 10 \qquad p2 := 6 \qquad p3 := 4, 4.01 .. 12$

Thus, the unit profit of product 3 should be $5.63 before it becomes economical to produce.

c. To analyze this, we need to rewrite the objective function and change the constraints (unfortunately, we are not allowed to have parameters in the constraints):

$f(x) := 10 \cdot x_1 + 6 \cdot x_2 + 4 \cdot x_3 \qquad x := \begin{pmatrix} 0 \\ 0 \\ 0 \end{pmatrix}$

Given

$\qquad x_1 + 2 \cdot x_2 + x_3 \le 100$

$\qquad 10 \cdot x_1 + 4 \cdot x_2 + 5 \cdot x_3 \le 600 \cdot 1.1 \qquad$ (10% increase)

$\qquad 3 \cdot x_1 + 2 \cdot x_2 + x_3 \le 300$

$\qquad x_1 \ge 0 \quad x_2 \ge 0 \quad x_3 \ge 0$

$x := \text{Maximize}(f, x)$

$$x = \begin{pmatrix} 57.5 \\ 21.25 \\ 0 \end{pmatrix}$$

Product 1: $x_1 = 0$

Product 2: $x_2 = 21.25$　　Profit:　$PR2 := f(x)$

Product 3: $x_3 = 0$　　　　　　　　$PR2 = 702.5$　$/day

PR2 − PR1 = 52.5　　It is not economical to schedule overtime, because the increase of profit is less than the additional labor cost needed ($60).

7.3 Non-Linear Programming

If any of the equalities and inequalities in the constraints and/or the objective function is non-linear, the optimization problem is referred to as the Non-Linear Programming. *Maximize/minimize* function can also be used to solve this type of problem. Procedure 7.6 demonstrates the use of these functions, which is similar to that discussed in the previous sections.

Procedure 7.6: *minimize/maximize for Non-Linear Programming*

Suppose we want to solve the following optimization problem:

Maximize　　$f(x) := 10 \cdot x_1 + 4.4 \cdot (x_2)^2 + 2 \cdot x_3$

Subject to　　$x_1 \geq 2$　$x_2 \geq 0$　$x_3 \geq 0$　　$(x_2)^2 + 0.5 \cdot (x_3)^2 \geq 3$

　　　　　　　　$x_1 + 4 \cdot x_2 + 5 \cdot x_3 \leq 45$　　　　$x_1 + 3 \cdot x_2 + 2 \cdot x_3 \leq 40$

1. Set up the optimization problem:

ORIGIN := 1

$f(x) := 10 \cdot x_1 + 4.4 \cdot (x_2)^2 + 2 \cdot x_3$

Given

　　$x_1 \geq 2$　$x_2 \geq 0$　$x_3 \geq 0$　　$(x_2)^2 + 0.5 \cdot (x_3)^2 \geq 3$

　　$x_1 + 4 \cdot x_2 + 5 \cdot x_3 \leq 45$　　　$x_1 + 3 \cdot x_2 + 2 \cdot x_3 \leq 40$

2. Create a function to obtain the optimum point by using *maximize* function:

$p(x) := Maximize(f, x)$　　　The argument x is a vector for the initial guesses

3. Calculate the optimum points p(x) for several sets of initial guesses and calculate the objective function to determine the global maximum:

$$x := \begin{pmatrix} 20 \\ 1 \\ 0 \end{pmatrix} \qquad p(x) = \begin{pmatrix} 34.804 \\ 1.732 \\ 0 \end{pmatrix} \qquad f(p(x)) = 361.238$$

Note that for non-linear programming, the *maximize* (or *minimize*) function might give us the local optima.

$$x := \begin{pmatrix} 2 \\ 1 \\ 0 \end{pmatrix} \qquad p(x) = \begin{pmatrix} 2 \\ 10.75 \\ 0 \end{pmatrix} \qquad f(p(x)) = 528.475$$
　　　　　　　　　　　　　　　　　　　　(maximum)

$$x := \begin{pmatrix} 10 \\ 0 \\ 1 \end{pmatrix} \qquad p(x) = \begin{pmatrix} 32.753 \\ 0 \\ 2.449 \end{pmatrix} \qquad f(p(x)) = 332.425$$

Example set 7.3

1. A chemical company sells 3 products: A, B, and C. Its revenue function can be represented as:

$$R(x) := 10 \cdot x_1 + 4 \cdot \left(x_2\right)^2 + 5.5 \cdot x_3 \qquad \$ \ 10^4/\text{month}$$

where x_1, x_2, and x_3 are the monthly production rates of A, B, and C. From the breakeven analysis, the minimum production rates of A, B, and C must be 5, 2, and 1 ton/month, respectively. The production rates of B and C must also follow the following restriction:

$$2 \cdot \left(x_2\right)^2 + 3 \cdot \left(x_3\right)^2 \geq 15$$

Additional restrictions must be considered due to the raw material availability:

$$x_1 + 4 \cdot x_2 + 8 \cdot x_3 \leq 40 \qquad x_1 + 3 \cdot x_2 + x_3 \leq 35$$

For maximum revenue, determine the production rate per month for each chemical. What is the revenue at this optimum points?

Solution: ORIGIN:= 1

$$R(x) := 10 \cdot x_1 + 4 \cdot \left(x_2\right)^2 + 5.5 \cdot x_3$$

Given

$$x_1 \geq 5 \qquad x_2 \geq 2 \qquad x_3 \geq 1 \qquad 2 \cdot \left(x_2\right)^2 + 3 \cdot \left(x_3\right)^2 \geq 15 \qquad \text{(production rate)}$$

$$x_1 + 4 \cdot x_2 + 8 \cdot x_3 \leq 40 \qquad\qquad\qquad\qquad \text{(material availability)}$$

$$x_1 + 3 \cdot x_2 + x_3 \leq 35 \qquad\qquad\qquad\qquad \text{(material availability)}$$

p(x) := Maximize(R, x)

$$x := \begin{pmatrix} 2 \\ 2 \\ 0 \end{pmatrix} \qquad p(x) = \begin{pmatrix} 22.202 \\ 2.449 \\ 1 \end{pmatrix} \qquad R(p(x)) = 251.52$$

$$x := \begin{pmatrix} 2 \\ 1 \\ 2 \end{pmatrix} \qquad p(x) = \begin{pmatrix} 5 \\ 6.75 \\ 1 \end{pmatrix} \qquad R(p(x)) = 237.75$$

Thus, the optimum production rates are 22.20 ton A/month, 2.45 ton B/month, and 1.00 ton C/month with a revenue of $251.52 10^4/month.

2. Thygeson and Grossmann [AICHE J., 16, 749 (1970)] studied the optimal design of a through circulation system for drying catalyst pellets. To maximize the production rate, the following optimization problem was obtained:

Minimize $\qquad f(x) := 0.0064 x_1 \cdot \left[\exp\left[-0.184 \left(x_1\right)^{0.3} \cdot x_2 \right] - 1 \right]$

Subject to $\qquad 1.2 \cdot 10^{13} - \left(3000 + x_1\right) \cdot \left(x_1\right)^2 \cdot x_2 \geq 0$

$$4.1 - \exp\left[0.184 \left(x_1\right)^{0.3} \cdot x_2 \right] \geq 0$$

$$x_1 \geq 0 \qquad x_2 \geq 0$$

where x_1 is the fluid velocity and x_2 is the bed depth. Determine the design fluid velocity and the bed depth that give maximum production rate.

Solution: ORIGIN:= 1

$$f(x) := 0.0064 x_1 \cdot \left[\exp\left[-0.184 \left(x_1\right)^{0.3} \cdot x_2 \right] - 1 \right]$$

Given

$$1.2 \cdot 10^{13} - \left(3000 + x_1\right) \cdot \left(x_1\right)^2 \cdot x_2 \geq 0 \qquad x_1 \geq 0$$

$$4.1 - \exp\left[0.184 \left(x_1\right)^{0.3} \cdot x_2 \right] \geq 0 \qquad x_2 \geq 0$$

p(x) := Minimize(f, x)

$$x := \begin{pmatrix} 100 \\ 100 \end{pmatrix} \qquad p(x) = \begin{pmatrix} 77136.322 \\ 0.024 \end{pmatrix} \qquad f(p(x)) = -59.437$$

$$x := \begin{pmatrix} 1 \\ 1 \end{pmatrix} \qquad p(x) = \begin{pmatrix} 31765.577 \\ 0.342 \end{pmatrix} \qquad f(p(x)) = -153.714$$

$$x := \begin{pmatrix} 10 \\ 100 \end{pmatrix} \qquad p(x) = \begin{pmatrix} 495.587 \\ 1.097 \end{pmatrix} \qquad f(p(x)) = -2.307$$

$$x := \begin{pmatrix} 0 \\ 0 \end{pmatrix} \qquad p(x) = \begin{pmatrix} 0 \\ 0 \end{pmatrix} \qquad f(p(x)) = 0$$

$$x := \begin{pmatrix} 100 \\ 50 \end{pmatrix} \qquad p(x) = \begin{pmatrix} 5095.947 \\ 0.533 \end{pmatrix} \qquad f(p(x)) = -23.442$$

$$x := \begin{pmatrix} 10 \\ 50 \end{pmatrix} \qquad p(x) = \begin{pmatrix} 84920.92 \\ 0.019 \end{pmatrix} \qquad f(p(x)) = -53.564$$

$$x := \begin{pmatrix} 1 \\ 70 \end{pmatrix} \qquad p(x) = \begin{pmatrix} 0.7 \\ 0 \end{pmatrix} \qquad f(p(x)) = 0$$

Unfortunately, in this case *minimize* function gives us many solutions depending on the initial guesses, so that we are not certain whether -153.714 is the minimum.

This might be due to scaling problem, one variable is several order of magnitude higher than the other. If we have any idea about the order of magnitude of the variables, we had better rescale the variables so that they take on values between 0.1 and 10.

For this problem, it is known that the fluid velocity is in the order of tens of thousands while the bed depth is in the order of one tenth to 10. By substituting $x_{1(old)} = 10^4 x_{1(new)}$, the objective function and constraints become:

$$f(x) := 64 \cdot (x_1) \cdot \left[\exp\left[-0.184\left(10^4 \cdot x_1 \right)^{0.3} \cdot x_2 \right] - 1 \right]$$

Given

$$120 - \left(3 + 10 \cdot x_1 \right) \cdot \left(x_1 \right)^2 \cdot x_2 \geq 0 \qquad\qquad x_1 \geq 0$$

$$4.1 - \exp\left[0.184\left(10^4 \cdot x_1 \right)^{0.3} \cdot x_2 \right] \geq 0 \qquad\qquad x_2 \geq 0$$

$$p(x) := \text{Minimize}(f, x)$$

Since now we know the order of magnitude of the variables, i.e., between 0.1 to 10, we can try some initial gueses in this range:

$$x := \begin{pmatrix} 1 \\ 1 \end{pmatrix} \qquad p(x) = \begin{pmatrix} 3.177 \\ 0.342 \end{pmatrix} \qquad f(p(x)) = -153.714$$

$$x := \begin{pmatrix} 10 \\ 0 \end{pmatrix} \qquad p(x) = \begin{pmatrix} 3.177 \\ 0.342 \end{pmatrix} \qquad f(p(x)) = -153.714$$

$$x := \begin{pmatrix} 5 \\ 10 \end{pmatrix} \qquad p(x) = \begin{pmatrix} 0 \\ 0 \end{pmatrix} \qquad f(p(x)) = 0$$

$$x := \begin{pmatrix} 2 \\ 3 \end{pmatrix} \qquad p(x) = \begin{pmatrix} 3.177 \\ 0.342 \end{pmatrix} \qquad f(p(x)) = -153.714$$

$$x := \begin{pmatrix} 0.5 \\ 5 \end{pmatrix} \qquad p(x) = \begin{pmatrix} 3.177 \\ 0.342 \end{pmatrix} \qquad f(p(x)) = -153.714$$

Thus, the solution is less sensitive to the initial guesses.

The optimum fluid velocity is 3.177×10^4 and the optimum bed depth is 0.342.

Problems

1. Substance A reacts according to an elementary autocatalytic reaction:
$$A + R \rightarrow R + R \qquad -r_A = kC_{A0}^2 X_A(1 - X_A)$$
where $-r_A$ is the rate of reaction, k is the reaction constant [= 1 liter/(mol.min)], C_{A0} is the concentration of A in the feed [= 1 mol/liter], and X_A is the conversion of A.

The process is carried out in a recycle reactor, which has the following design equation:
$$\tau = C_{A0}(R+1) \int_{\frac{RX_{Af}}{R+1}}^{X_{Af}} \frac{1}{-r_A} dX_A$$
where τ is the space time (residence time), R is the recycle ratio, and X_{Af} is the final conversion of A [= 0.9]. Determine the optimum recycle ratio that gives minimum reactor volume (minimum space time). What is the residence time of the reacting mixture in the reactor at this optimum recycle ratio?

2. Consider a gas pipeline transmission system where compressor stations are placed L miles apart [*Sherwood, A Course in Process Design*, MIT Press, Cambridge, 1963]. The total annual cost of this transmission system and its operation is
$$C(D, P_1, L, r) = 7.84D^2 P_1 + 450,000 + 36,900D$$
$$+ \frac{6.57 \times 10^6}{L} + \frac{772 \times 10^6}{L}\left(r^{0.219} - 1\right)$$
where D is the pipe inside diameter (in), P_1 is the compressor discharge pressure (psia), L is the length between compressor stations (miles), r is the compression ratio (P_1/P_2), and P_2 is the compressor inlet pressure (psia). If the flow rate is 100×10^6 scf/day, determine the design variables, i.e., D, P_1, L, and r, for minimum annual cost.

The flow rate (Q) is related to the design variables:
$$Q = 3.39\sqrt{\frac{\left(P_1^2 - P_2^2\right)D^5}{fL}} \qquad \text{(scf/hr)}$$
where f is the friction factor ($= 0.008D^{-1/3}$).

3. Product R will be produced at 80 mol/hr (F_R) in a mixed flow reactor, in which the following first order reaction takes place:

$$A \to R \qquad r_R = kC_A$$

where r_R is the reaction rate of R formation, k is the reaction constant (= 0.5 hr^{-1}), and C_A is the concentration of A in the reactor. The concentration of reactant A in the feed (C_{A0}) is 0.2 mol/liter. The cost of reactant (P_s) at this concentration is \$0.80/mol A while the cost of reactor including installation, instrumentation, depreciation, etc. (P_r), is \$0.02/(hr·liter). Determine the reactor size, feed rate, and conversion that should be used for optimum operations, i.e., minimum cost. Also, determine the unit cost of R for these conditions. Assume that the unreacted A is discarded.

On an hourly basis, the total cost is

$$P_t = VP_r + F_{A0}P_s$$

where V is the volume of the reactor and F_{A0} is the molar flow rate of A. For a first-order reaction, the volume of reactor, which is derived from mole balance, is given by

$$V = \frac{F_{A0}X_A}{kC_{A0}(1 - X_A)}$$

where X_A is the conversion of A, from which we can also relate the product rate to the feed rate: $F_R = F_{A0}X_A$

Note that if the conversion is low, the reactor volume needed will be small, but the amount of reactant A needed for a certain product rate becomes large. On the other hand, if the conversion is high, the amount of reactant A will be small, but the rector volume needed will be large.

4. A company produces alcohol in two grades through a single process stream. The unit profit on product A is \$1.5/gallon and on product B is \$0.9/gallon. Since the purity of product A is higher than that of product B, product A needs 1.5 times the processing times of product B. Product B alone can be produced at 1500 gal/day. Contract sales require that at least 300 gal/day of product B be produced. For optimum profit, determine the amount of products A and B that should be produced per day. What is the profit?

5. In a three-stage compressor, the process of which is adiabatic reversible, one needs to optimize the discharge pressures from the first and the second stages so that the total work needed is minimized. For a given diatomic gas and final pressure (P_3 = 78 bar), the optimal pressures are obtained as the solution of the following problem:

$$\text{Minimize} \qquad f(P_1, P_2) = P_1^{\frac{2}{7}} + \left(\frac{P_2}{P_1}\right)^{\frac{2}{7}} + \left(\frac{P_3}{P_2}\right)^{\frac{2}{7}}$$

Subject to $\quad P_1 \geq 1$ (the gas enters the three-stage compressor at 1 bar)

$\quad P_1 \leq P_2 \leq P_3$ (the gas pressures are monotonically increasing from inlet to outlet)

where P_1 and P_2 are the discharge pressures from the first and the second stages, respectively. Determine the optimal discharge pressures from the first and the second stages.

6. The total heat transfer area of three cross-flow heat exchangers in series, as shown in the figure below, is to be minimized. Cold fluid is heated in each heat exchanger by hot fluid. Each heat exchanger can be modeled by a steady-state lumped model:

$$WC_p(T_i - T_{i-1}) = U_i A_i(t_i - T_i)$$

where WC_p = heat capacity of the cold fluid $[= 10^5 \text{ Btu/(hr°F)}]$
T_i = temperature of cold fluid leaving heat exchanger i
T_{i-1} = temperature of cold fluid entering heat exchanger i
U_i = overall heat transfer coefficient in heat exchanger i
A_i = heat transfer area of heat exchanger i
t_i = temperature of hot fluid entering heat exchanger i

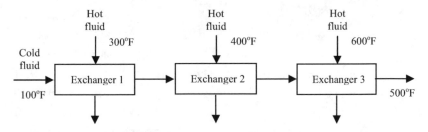

a. What are the design variables that can be adjusted to minimize the total heat transfer area of this three-stage heat exchanger?
b. Write the objective function, i.e., the total heat transfer area as a function of the design variables.
c. Write the constraints considering that the temperatures of the cold fluid are monotonically increasing from inlet to outlet and the temperature of the hot fluid entering a heat exchanger must be higher than that of the cold fluid leaving the heat exchanger.

d. Determine the optimum design variables and the minimum total heat transfer area of the system.

Data: $U_1 = 120$ Btu/(hr·ft^2·$^\circ$F)

$\quad\quad U_2 = 80$ Btu/(hr·ft^2·$^\circ$F)

$\quad\quad U_3 = 40$ Btu/(hr·ft^2·$^\circ$F)

7. In a pharmaceutical industry, three highly demanded products (A, B, and C) are produced in a single process stream consisting three consecutive steps. It is possible to run the equipment in each step 24 hours a day. Contract sales require that at least 2 kg/day of product A be produced.

	Product A	Product B	Product C
Profit, $/kg	40	60	80
Time needed to process one kg of product in every step, hours			
Step 1	3	2	4
Step 2	1	1	7
Step 3	1	3	3

a. How many kilograms per day should each product be produced to give maximum profit? What is the maximum profit?
b. Is there unused time available on any of the steps with the optimal solution? If yes, which step does not run 24 hours a day?
c. Unfortunately, there is an energy crisis leading to widespread blackouts. A few power generators with limited capacity are available in the production site, but one of them cannot run for more than 12 hours/day. As a consequence, one of the production steps must not run more than 12 hours a day. As a production manager, decide which step should operate no more than 12 hours. Do some calculations to support your reasoning.

Chapter 8
Differential Equation

Differential equations are of fundamental importance in engineering. For chemical engineering, differential equations appear for example in transport phenomena, chemical reaction engineering, unit operations, and process control. Although Mathcad has a limited capability in solving differential equations, it can solve many problems in chemical engineering. We will discuss ordinary differential equations first followed by partial differential equation. Keep in mind that all of the functions used for solving differential equations require dimensionless variables.

8.1 Ordinary Differential Equation (ODE)

An ordinary differential equation is a differential equation that has only one independent variable. An ODE is called an n-th order ODE if the highest derivative in the equation is of order n and is called linear if it is linear in the dependent variable and all of its derivatives.

A differential equation can also be classified according to the known conditions. A differential equation with initial conditions is called an initial value problem (IVP). In IVP, the conditions (the value of the dependent variable and/or all of its derivatives) are known at a single value of independent variable, which is the starting point. If the conditions are given at two different values of independent variable, the problem is called boundary value problem (BVP). The number of conditions in IVP or BVP must be the same as the order of ODE.

8.1.1. First Order ODE

To solve a first order ODE, in Mathcad we can use several functions, such as *odesolve*, *rkfixed*, and *Rkadapt* functions, which will be discussed in detail. The ODE must be linear in its highest derivative.

For *odesolve* function, by default, it uses **Adams/BDF** method, which means **Adams** (Adams-Bashford) or **BDF** (Backward Differentiation Formula) method depending on whether the system is non-stiff or stiff. The other available methods can be accessed by right clicking on *odesolve* and choose a particular method from the pop-up menu. Procedure 8.1 shows how to use *odesolve* function to solve a first order ODE.

Procedure 8.1: *odesolve for a single first order ODE*

Suppose we want to solve a single ODE from x = 2 to x = 10:

$$\frac{d}{dx}y - 0.02 \cdot y = -x \qquad \text{with an initial condition:} \qquad y(2) = 3$$

1. Define the starting and the end (terminal) points of the independent variable x:

  Starting point: $a := 2$

  End point: $b := 10$

  The end point can be greater or less than the starting point.

2. Set up the ODE along with the initial condition in a *Given* block

  Given

$$\frac{d}{dx}y(x) - 0.02 \cdot y(x) = -x$$

$$y(a) = 3$$

$$y := Odesolve\,(x, b)$$

  Note: a. The dependent variable (y) must be typed as a function
  b. In a Given block, we must use the Boolean equal sign (bold equal sign).
  c. The derivative must be typed on the left side of the equation.

  odesolve function can be typed any way we want, such as *Odesolve* or *ODEsolve*.

If we want to set the number of points, the syntax is Odesolve(x,b,npoints). Remember that the smaller the number of points, the less accurate the interpolated value will be.

The solution y is a function that can be plotted as a function of the independent variable x:

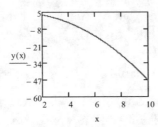

It can also be used to calculate the dependent variable at a certain value of the independent variable:

For example at x = 5:

$$y(5) = -7.59$$

If the ODE contains a parameter, we have a parametric problem and the following procedure should be followed:

  Given

$$\frac{d}{dx}y(x) - p \cdot y(x) = -x \qquad y(a) = 3$$

$$y(p) := Odesolve\,(x, b) \qquad \text{The parameter } p \text{ is now included as an argument.}$$

To plot the solution for a certain parameter, it is necessary to define the independent variable as a range variable to reduce the time needed for plotting:

$$x := 2, 2.05 .. 10$$

For p = 0.02 at x = 5: $y(0.02, 5) = -7.59$

Note that the parameter must be typed **before** the independent variable.

Since *odesolve* function uses a numerical method, the solution obtained is an approximation. The values of the dependent variable (i.e., the solution) are saved at equally spaced discrete points of the independent variable. By default, the number of points is 1000. If the value of the dependent variable is needed at an independent variable other than the discrete points, the solution will be interpolated using *lspline* function (see Section 5.1.2).

For *rkfixed* function, as suggested by its name, it uses a **Fixed** step Runge-Kutta method to solve the ODE. In some applications, for example, when the rate of change of the dependent variable is large in a certain region, we may need to use an adaptive method, which uses different step size in different region. In such cases, we could use *Rkadapt* function. By using *rkfixed/Rkadapt* function, we can also solve an ODE with decreasing independent variable, i.e., the starting point is larger than the end point. Procedure 8.2 demonstrates the use of *rkfixed* function while Procedure 8.3 demonstrates the use of *Rkadapt* function.

Procedure 8.2: *rkfixed* for a first order ODE

Suppose we want to solve a single ODE from x = 2 to x = 10:

$$\frac{d}{dx}y - 0.02y = -x \qquad \text{with an initial condition:} \qquad y(2) = 3$$

1. Set up the differential equation:

$D(x, y) := 0.02y - x$ Here, D represents the derivative function

2. Set up the initial condition:

$yi := 3$

3. Solve the ODE using *rkfixed* function:

$x0 := 2$ the starting point

$n := 100$ the number of interval between x0 and x

$u(x) := rkfixed(yi, x0, x, n, D)$ Note that x here is the end point. Of course we can in fact set the end point and remove the argument, but this approach is recommended.

	0	1
0	2	3
1	2.08	2.841
2	2.16	2.676
3	2.24	...

$u(10) =$

u is a matrix containing the values of the independent variable x in the first column $(u^{\langle 0 \rangle})$ and the values of dependent variable y in the second column $(u^{\langle 1 \rangle})$.

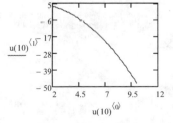

The dependent variable can be plotted as a function of the independent variable

4. If we want to calculate the dependent variable at a certain value of the independent variable, set up a function for this purpose:

$$y(x) := \left(u(x)^{\langle 1 \rangle}\right)_n \qquad \text{We pick the last element of the second column.}$$

For example at x = 5: $y(5) = -7.59$

If the ODE contains a parameter, we have a parametric problem and the following procedure should be followed:

1. Set up the differential equation:

$$D(p, x, y) := p \cdot y - x \qquad$$ Now, p is included in the derivative function. Any additional parameters must be typed **before** the independent and dependent variables.

2. Set up the initial condition: $yi := 3$

3. Solve the ODE using *rkfixed* function:

$x0 := 2$ the starting point

$n := 100$ the number of interval between x0 and x

$u(p, x) := \text{rkfixed}(yi, x0, x, n, D(p))$ Note that x here is the end point and p has been included in the argument of D and u.

	0	1
0	2	3
1	2.08	...

$u(0.02, 10) =$

4. To calculate the dependent variable at a certain value of the independent variable:

$$y(p, x) := \left(u(p, x)^{\langle 1 \rangle}\right)_n \qquad \text{We pick the last element of the second column.}$$

For example for p = 0.02 at x = 5: $y(0.02, 5) = -7.59$

Procedure 8.3: *Rkadapt* for a first order ODE

Suppose we want to solve a single ODE from x = 2 to x = 10:

$$\frac{d}{dx} y = \frac{1}{3.98 - x^2} \qquad \text{with an initial condition:} \quad y(2) = 2$$

This ODE has an analytical solution: $y_a(x) := 2 + \int_2^x \frac{1}{3.98 - x^2} \, dx$

1. Set up the initial condition:

$yi := 2$

2. Set up the differential equation:

$$D(x, y) := \frac{1}{3.98 - x^2}$$

3. Solve the ODE using *Rkadapt* function:

$x0 := 2$ the starting point

$n := 30$ the number of interval between x0 and x

$u(x) := \text{Rkadapt}(yi, x0, x, n, D)$ x is the end point

Note: Make sure capital R is used in *Rkadapt* function.

	0	1
0	2	2
1	2.267	0.937
2	2.533	0.78
3	2.8	...

$u(10) =$

u is a matrix containing the values of the independent variable x in the first column $(u^{\langle 0 \rangle})$ and the values of dependent variable y in the second column $(u^{\langle 1 \rangle})$.

If we want to solve it using *rkfixed* function, then

$$w := \text{rkfixed}(yi, 2, 10, 30, D) \qquad$$ Note that here the number of intervals used is 30.

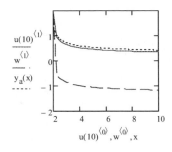

$y_a(x)$ is the analytical solution.

Thus, the results obtained using *rkfixed* function (dashed line) with only 30 intervals is incorrect. In order to get the correct results, the number of intervals must be increased.

On the other hand, *Rkadapt* function does a great job with a fewer number of intervals.

4. If we want to calculate the dependent variable at a certain value of the independent variable, set up a function for this purpose:

$$y(x) := \left(u(x)^{\langle 1 \rangle}\right)_n$$

For example at x = 5:

$$y(5) = 0.505$$

8.1.2. System of First Order ODEs

We can also use the *odesolve*, *rkfixed*, and *Rkadapt* functions to solve a system of first order ODEs. The advantage of using *rkfixed /Rkadapt* function to solve a system of first order ODEs is that we can solve parametric problems by including the parameters in the arguments of the derivative function; *odesolve* function cannot solve a system of first order ODEs with parameters. Procedures 8.4 and 8.5 show the use of *odesolve* and *rkfixed* functions, respectively, for this purpose.

Procedure 8.4: *odesolve* for a system of ODEs

Suppose we want to solve two simultaneous ODEs from x = 2 to x = 10:

$$\frac{d}{dx}y1 - 0.02 \cdot y1 = -x$$

$$\frac{d}{dx}y2 + x \cdot y2 = y1$$

with initial conditions: $y1(2) = 3$ $y2(2) = 0$

1. Define the starting and the end (terminal) points of the independent variable x:

    Starting point: $a := 2$

    End point: $b := 10$

2. Set up the ODE along with the initial condition in a *Given* block

    Given
    $$\frac{d}{dx}y1(x) - 0.02 \cdot y1(x) = -x \qquad y1(a) = 3$$

    $$\frac{d}{dx}y2(x) + x \cdot y2(x) = y1(x) \qquad y2(a) = 0$$

3. Solve the ODE using *odesolve* function:

$$\begin{pmatrix} y1 \\ y2 \end{pmatrix} := \text{Odesolve}\left[\begin{pmatrix} y1 \\ y2 \end{pmatrix}, x, b\right]$$

Note: a. The first argument in the *odesolve* function here is a vector containing the name of the functions that we want to solve.

b. The *odesolve* function in this case returns a vector of functions of x, which is the solution to the system of ODEs.

The solutions y1 and y2 are functions that can be plotted:

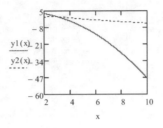

They can also be used to calculate the dependent variables at a certain value of the independent variable:

For example at x = 5:

$$y1(5) = -7.59$$
$$y2(5) = -1.362$$

Procedure 8.5: *rkfixed/Rkadapt* *for a system of ODEs*

Suppose we want to solve two simultaneous ODEs from x = 2 to x = 10:

$$\frac{d}{dx}y0 = 0.02 \cdot y0 - x \qquad \frac{d}{dx}y1 = -x \cdot y1 + y0$$

with initial conditions: $y0(2) = 3$ $y1(2) = 0$

1. Set up the initial conditions:

$$vyi := \begin{pmatrix} 3 \\ 0 \end{pmatrix}$$

Note that the conditions are entered in a vector

2. Set up a derivative function containing the differential equations (y is an array):

$$D(x,y) := \begin{pmatrix} 0.02 \cdot y_0 - x \\ -x \cdot y_1 + y_0 \end{pmatrix}$$

Thus, D is a vector function. The first element is for dy0/dx and the second is for dy1/dx.

3. Solve the ODEs using *rkfixed (Rkadapt)* function:

$$x0 := 2$$ the starting point

$$n := 100$$ the number of interval between x0 and x

$$u(x) := rkfixed(vyi, x0, x, n, D)$$ x is the end point

	0	1	2
0	2	3	0
1	2.08	2.841	0.215
2	2.16	2.676	0.385
3	2.24	2.504	...

$u(10) =$ (at left of table)

u is a matrix containing the values of the independent variable x in the first column and the values of dependent variables y0 and y1 in the second and third columns, respectively.

Below is the plot of y0 and y1 vs. x:

$$X(x) := u(x)^{\langle 0 \rangle} \qquad Y0(x) := u(x)^{\langle 1 \rangle} \qquad Y1(x) := u(x)^{\langle 2 \rangle}$$

If we want to calculate the dependent variables at a certain value of the independent variable, set up the functions for this purpose:

$$y0(x) := Y0(x)_n \qquad y1(x) := Y1(x)_n$$

For example at x = 5:

$$y0(5) = -7.59 \qquad y1(5) = -1.362$$

8.1.3. Higher Order ODE – Initial Value Problem

Higher order IVP ODEs, including parametric problems, can be solved using *odesolve*, *rkfixed*, or *Rkadapt* function. For *odesolve* function, we can directly use the function to solve the problem without transforming the ODE into a system of first order ODEs. For *rkfixed/Rkadapt* function, however, we must transform the higher order

ODE into a system of first order ODEs. Procedures 8.6 and 8.7 show the use of *odesolve* and *rkfixed* functions, respectively, for this purpose.

Procedure 8.6: *odesolve for a higher order ODE (IVP)*

Suppose we want to solve a third order ODE from t = 0 to t = 10:

$$2 \cdot \frac{d^3}{dt^3}y - \frac{d^2}{dt^2}y + 3 \cdot y = 1 + \sin(2 \cdot t)$$

with initial conditions at t = 0: $y = 0$ $\frac{d}{dt}y = 0$ $\frac{d^2}{dt^2}y = 0$

1. Define the starting and end points of the independent variable: $a := 0$ $b := 10$

2. Set up the ODE along with the initial conditions in a *Given* block:

Given

$$2 \cdot \frac{d^3}{dt^3}y(t) - \frac{d^2}{dt^2}y(t) + 3 \cdot y(t) = 1 + \sin(2 \cdot t)$$

The derivative of the highest order must be typed on the left side of the equation.

$$y(a) = 0 \qquad y'(a) = 0 \qquad y''(a) = 0$$

The derivatives in the initial conditions must be represented by y', y", etc.

3. Solve the ODE using *odesolve* function

$$y := Odesolve(t, b)$$

Plot of y as a function of t:

Press [Ctrl F7] to type the prime and [Ctrl F7] twice to type the double prime.

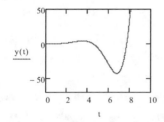

To calculate y at a certain value of t, for example at t = 7:

$$y(7) = -41.287$$

Procedure 8.7: *rkfixed/Rkadapt for a high order ODE (IVP)*

Suppose we want to solve a third order ODE from t = 0 to t = 10:

$$2 \cdot \frac{d^3}{dt^3}y - \frac{d^2}{dt^2}y + 3 \cdot y = 1 + \sin(2 \cdot t)$$

with initial conditions at t = 0: $y = 0$ $\frac{d}{dt}y = 0$ $\frac{d^2}{dt^2}y = 0$

We can transform this third order ODE into 3 first order ODEs:

$$\frac{d}{dt}y0 = y1$$

$$\frac{d}{dt}y1 = y2$$

$$2 \cdot \frac{d}{dt}y2 - y2 + 3 \cdot y0 = 1 + \sin(2 \cdot t)$$

1. Set up the initial conditions:

$$vyi := \begin{pmatrix} 0 \\ 0 \\ 0 \end{pmatrix}$$

Note that the conditions are entered in a vector. The first element is for y0, the second is for y1, and the third is for y2

2. Set up a derivative function containing the differential equations (y is an array):

$$D(t, y) := \begin{bmatrix} y_1 \\ y_2 \\ \frac{1}{2} \cdot \left(y_2 - 3 \cdot y_0 + 1 + \sin(2 \cdot t) \right) \end{bmatrix}$$

Thus, D is a vector function. The first element is for dy0/dt, the second is for dy1/dt, and the third is for dy2/dt.

3. Solve the ODEs using *rkfixed (Rkadapt)* function:

$t0 := 0$ $n := 100$ $u(t) := Rkadapt(vyi, t0, t, n, D)$ t is the end point

		0	1	2	3
u(10) =	0	0	0	0	0
	1	0.1	$8.859 \cdot 10^{-5}$	$2.711 \cdot 10^{-3}$	0.056
	2	0.2	$7.513 \cdot 10^{-4}$	0.012	0.126
	3	0.3	$2.68 \cdot 10^{-3}$	0.028	0.207

The matrix columns contain the values of t, y0, y1, and y2, respectively.

The plot of y0 vs. t:

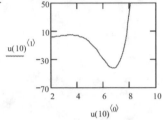

4. To calculate y at a certain value of t, for example at t = 7:

$$y0(t) := \left(u(t)^{\langle 1 \rangle} \right)_n \qquad y0(7) = -41.287$$

8.1.4. Higher Order ODE – Boundary Value Problem

Boundary value problems, including parametric problems, can also be solved using *odesolve, rkfixed*, or *Rkadapt* function. For *odesolve* function, as in IVP, the constraint should be given in the form of *y(a)=b* or *y'(a)=b*. *Odesolve* function does not accept more complicated constraints such as *c·y'(a)+d·y(a)=b*. In this case, we may use *rkfixed* or *Rkadapt* function. The procedure how to use *odesolve* function for this purpose is the same as that for IVP problem, as shown in Procedure 8.8.

Before we can use *rkfixed* or *Rkadapt* function, which in fact can only solve IVP, we must use *sbval* function to find the unspecified (unknown) initial values that match the other end values. After we find the unknown initial values, the BVP reduces to IVP. Procedure 8.9 shows how to use *sbval* and *rkfixed/Rkadapt* functions for BVP.

Procedure 8.8: *odesolve for a second order ODE (BVP)*

Suppose we want to solve a second order ODE from t = 0 to t = 5:

$$\frac{d^2}{dt^2}y + 3 \cdot \left(\frac{d}{dt}y \right) - y = 5 \cdot t \quad \text{with boundary conditions: at t = 0:} \quad \frac{d}{dt}y = 0$$

$$\text{and at t = 5:} \quad y = 2$$

1. Define the starting and end points of the independent variable:

    $a := 0$ $b := 5$

2. Set up the ODE along with the initial conditions in a *Given* block:

    Given

$$\frac{d^2}{dt^2}y(t) + 3 \cdot \left(\frac{d}{dt}y(t) \right) - y(t) = 5 \cdot t \qquad y'(a) = 0 \qquad y(b) = 2$$

3. Solve the ODE using *odesolve* function

$$y := \text{Odesolve}(t, b)$$

The solution y is a function that can be plotted as a function of the independent variable t

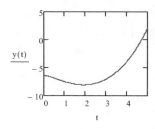

It can also be used to calculate the dependent variable at a certain value of the independent variable:

For example at t = 3:

$$y(3) = -7.078$$

Procedure 8.9: *sbval & rkfixed/Rkadapt*

Suppose we want to solve a second order ODE from t = 0 to t = 5:

$$\frac{d^2}{dt^2}y + 3 \cdot \left(\frac{d}{dt}y\right) - y = 5 \cdot t$$

with boundary conditions: at t = 0: $\frac{d}{dt}y = 0$ and at t = 5: $y = 2$

We can transform this ODE into 2 first order ODEs:

$$\frac{d}{dt}y0 = y1 \qquad \frac{d}{dt}y1 + 3 \cdot y1 - y0 = 5 \cdot t$$

with boundary conditions: at t = 0: $y1 = 0$ and at t = 5: $y0 = 2$

1. Define the starting and end points of the independent variable:

$$a := 0 \qquad\qquad b := 5$$

2. Create a vector containing the intial guess(es) of the unknown condition(s) at a:

$$vi_0 := 1$$ In this case, we have only one unknown condition at a, which must be assigned to an element of a vector.

3. Set up the differential equations: $D(t, y) := \begin{pmatrix} y_1 \\ -3 \cdot y_1 + y_0 + 5 \cdot t \end{pmatrix}$

4. Create the IV vector containing the known & unknown conditions at a (initial values):

$$IV(a, vi) := \begin{pmatrix} vi_0 \\ 0 \end{pmatrix}$$ IV stands for Initial Value(s). Of course, other names are fine. The first element is for y0, the initial value of which is unknown, and the second is for y1, the initial value of which is known.

5. Create the EV vector containing the known condition(s) at b:

$$EV(b, y) := y_0 - 2$$ EV stands for End Value(s). The known end value is y0 = 2 and must be transformed to $y_0 - 2 = 0$

6. Obtain the unknown condition(s) at a using *sbval* function:

$$s := \text{sbval}(vi, a, b, D, IV, EV) = (-6.424)$$

7. Once we know the missing initial value(s), BVP reduces to IVP and we can solve the ODE using rkfixed/Rkadapt function:

$$n := 100 \qquad vi := \begin{pmatrix} s_0 \\ 0 \end{pmatrix} \qquad u := \text{rkfixed}(vi, a, b, n, D)$$

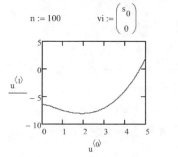

8.1.5. Second Order ODE – Cylindrical/Spherical Coordinate

Since second order ODE derived from energy or species balance in the cylindrical or spherical coordinate frequently occurs in chemical engineering and the Mathcad procedure to solve this type of ODE needs further attention, we discuss the issue separately in this section.

The general second order ODE obtained from steady-state one-dimensional (in the radial direction) energy or species balance is as follows:

$$\frac{1}{r}\frac{d}{dr}\left(r\frac{dY}{dr}\right) + f(Y) = 0 \tag{8.1}$$

for cylindrical coordinate, or

$$\frac{1}{r^2}\frac{d}{dr}\left(r^2\frac{dY}{dr}\right) + g(Y) = 0 \tag{8.2}$$

for spherical coordinate. In Equations (8.1) and (8.2), Y could be the temperature or concentration of a species in the system of interest, which is a function of r (the radial distance from the cylinder axis or from the center of sphere) and $f(Y)$ and $g(Y)$ are functions of Y representing the energy or species generation term.

In Mathcad, of course we can use *odesolve* or *rkfixed/Rkadapt* function to solve the problem. However, we need to expand the derivative term in Equation (8.1) or (8.2) to obtain:

$$\left(\frac{d^2Y}{dr^2} + \frac{1}{r}\frac{dY}{dr}\right) + f(Y) = 0 \tag{8.3}$$

for cylindrical coordinate, or

$$\left(\frac{d^2Y}{dr^2} + \frac{2}{r}\frac{dY}{dr}\right) + g(Y) = 0 \tag{8.4}$$

for spherical coordinate.

For *odesolve* function, the second order ODE in the form of Equation (8.3) or (8.4) is typed in the *Given* block. For *rkfixed/Rkadapt* function, of course we need to further transform the ODE to a system of first order ODEs. Because of the existence of r in the denominator of the second term on the left side of Equation (8.3) or (8.4), if any of the boundary conditions is known at $r = 0$ (at the center), which is very common, as an approximation we need to assume that the boundary condition is known at a very small r, say $r = 10^{-8}$.

8.1.6. System of Higher Order ODEs (IVP or BVP)

We can also use the *odesolve*, *rkfixed*, or *Rkadapt* function to solve a system of higher order ODEs. Again, although *odesolve* function is much simpler than *rkfixed/Rkadapt* function, it cannot solve a system of higher order ODEs with parameters or with complicated conditions. In addition, if the ODEs are very complex, *odesolve* function may not be able to solve the problem. In those cases, *rkfixed/Rkadapt* function can be used.

Procedures 8.10 and 8.11 show the use of *odesolve* and *rkfixed/Rkadapt* functions, respectively, for this purpose. In Procedure 8.11, we only demonstrate the procedure for solving BVP. For solving IVP, after transforming the problem to a system of first order ODEs, the procedure is exactly the same as that shown in Procedure 8.7.

Procedure 8.10: *odesolve* *for a system of higher order ODEs*

Suppose we want to solve the following system of ODEs from t = 0 to t = 5:

$$\frac{d^2}{dt^2}y0 + 3\cdot\left(\frac{d}{dt}y0\right) - y0 = 5\cdot t\cdot y1$$

$$\frac{d^2}{dt^2}y1 + 3\cdot\left(\frac{d}{dt}y1\right) - y1 = -2\cdot y0$$

with boundary conditions: at t = 0: $\quad \frac{d}{dt}y0 = 0 \qquad \frac{d}{dt}y1 = 1$

$$\text{and at t = 5:} \quad y0 = 2 \qquad y1 = 3$$

1. Define the starting and end points of the independent variable:

   $a := 0 \qquad\qquad b := 5$

2. Set up the ODEs along with the initial conditions in a *Given* block:

   Given

   $$\frac{d^2}{dt^2}y0(t) + 3\cdot\left(\frac{d}{dt}y0(t)\right) - y0(t) = 5\cdot t\cdot y1(t)$$

   $$\frac{d^2}{dt^2}y1(t) + 3\cdot\left(\frac{d}{dt}y1(t)\right) - y1(t) = -2\cdot y0(t)$$

   $y0'(a) = 0 \qquad y0(b) = 2 \qquad y1'(a) = 1 \qquad y1(b) = 3$

3. Solve the ODEs using *odesolve* function

   $$\begin{pmatrix} y0 \\ y1 \end{pmatrix} := \text{Odesolve}\left[\begin{pmatrix} y0 \\ y1 \end{pmatrix}, t, b\right]$$

Plot of y0 and y1 as functions of the independent variable t:

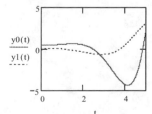

It can also be used to calculate the dependent variable at a certain value of the independent variable:

For example at t = 3:

$y0(3) = -1.183$

$y1(3) = -0.667$

Again, if the system of higher order ODEs is obtained from energy and species balances in the cylindrical or spherical coordinate, as described in Section 8.1.5, we need to expand the derivative term first before we can solve the problem. Also, if any of the boundary conditions is known at $r = 0$, we need to assume that the boundary condition is known at a very small r.

Procedure 8.11: sbval & rkfixed/Rkadapt *for a system of higher order ODEs*

Suppose we want to solve the following system of ODEs from t = 0 to t = 5:

$$\frac{d^2}{dt^2}f + 3\cdot\left(\frac{d}{dt}f\right) - f = 5\cdot t\cdot g \qquad \text{at } t = 0: \quad \frac{d}{dt}f = 0 \qquad \frac{d}{dt}g = 1$$

$$\frac{d^2}{dt^2}g + 3\cdot\left(\frac{d}{dt}g\right) - g = -2\cdot f \qquad \text{and at } t = 5: \quad f = 2 \qquad g = 3$$

We can transform these ODEs into 4 first order ODEs (with f = y0 and g = y2):

$$\frac{d}{dt}y0 = y1 \qquad \frac{d}{dt}y1 + 3\cdot y1 - y0 = 5\cdot t\cdot y2 \qquad \frac{d}{dt}y2 = y3 \qquad \frac{d}{dt}y3 + 3\cdot y3 - y2 = -2\cdot y0$$

with boundary conditions: at t = 0:   y1 = 0   y3 = 1   and at t = 5:   y0 = 2   y2 = 3

1. Define the starting and end points of the independent variable:

     a := 0   b := 5

2. Create a vector containing the intial guess(es) of the unknown condition(s) at a:

$$vi := \begin{pmatrix} 1 \\ 1 \end{pmatrix} \qquad\qquad\qquad \begin{pmatrix} y_1 \\ -3\cdot y_1 + y_0 + 5\cdot t\cdot y_2 \\ y_3 \\ -3\cdot y_3 + y_2 - 2\cdot y_0 \end{pmatrix}$$

3. Set up the differential equations: $D(t, y) :=$

4. Create the IV vector containing the known & unknown conditions at a (initial values) and the EV vector containing the known condition(s) at b:

$$IV(a, vi) := \begin{pmatrix} vi_0 \\ 0 \\ vi_1 \\ 1 \end{pmatrix} \qquad EV(b, y) := \begin{pmatrix} y_0 - 2 \\ y_2 - 3 \end{pmatrix}$$

5. Obtain the unknown condition(s) at a using *sbval* function:

$$s := sbval(vi, a, b, D, IV, EV) \qquad\qquad s = \begin{pmatrix} 0.419 \\ -0.122 \end{pmatrix}$$

Thus, the initial values: $vyi := \begin{pmatrix} s_0 \\ 0 \\ s_1 \\ 1 \end{pmatrix}$

6. BVP reduces to IVP and we can solve the ODE using rkfixed/Rkadapt function:

     n := 100   u := rkfixed(vyi, a, b, n, D)

Example set 8.1

1. To study the liquid level response, a tank with a cross-sectional area (A) of 2 m^2 is filled with water until the height of the liquid level (h) is 1.5 m. At t = 0, a valve below the tank is opened and the outlet flow rate is F_o. At the same time, 0.1 m ^{3}/min of water (F_i) enters the tank. If the outlet flow rate is a linear function of the liquid level:

$$F_o = \alpha \cdot h \qquad \alpha \text{ is a constant } (=0.2 \text{ m}^2/\text{min})$$

plot the height of the liquid in the tank as a function of time (t), say from t = 0 to t = 60 min and determine the height of the liquid level at t = 10 min.

The mass balance yields

$$A \cdot \frac{dh}{dt} = F_i - F_o$$

Solution: This is a first order ODE.

Known parameters: $A := 2$ $\alpha := 0.2$ $F_i := 0.1$ $t1 := 60$

Using *odesolve* function:

Given

$$h'(t) = \frac{F_i - \alpha \cdot h(t)}{A} \qquad h(0) = 1.5$$

$$h := Odesolve(t, t1)$$

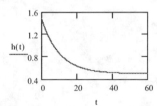

$$h(10) = 0.868 \quad m$$

Using *rkfixed* function:

Initial value: $h_0 := 1.5$ Set up the differential equation: $D(t, h) := \dfrac{F_i}{A} - \dfrac{\alpha}{A} \cdot h$

Solve the ODE using *rkfixed* function:

$$n := 100 \qquad u(t) := rkfixed(h, 0, t, n, D)$$

$$X(t) := u(t)^{\langle 0 \rangle} \qquad Y(t) := u(t)^{\langle 1 \rangle}$$

Height of liquid [m]

$$y(t) := Y(t)_n$$

$$y(10) = 0.868 \quad m$$

2. m-Xylene is produced in a plug flow reactor at 1500 R and 35 atm from mesitylene. Two reactions occur in the reactor [Fogler, *Elements of Chemical Reaction Engineering*, Prentice-Hall, Englewood Cliffs, 1992]:

   (1) Mesitylene (M) + Hydrogen (H) ------- m-Xylene (X) + Methane
   (2) m-Xylene (X) + Hydrogen (H) ------- Toluene (T) + Methane

The second reaction is not desirable because it will further convert m-xylene (the desired product) to toluene. The following system of ODEs can be set up from mol balance:

$$\frac{d}{d\tau}C_H = -k_1 \cdot C_H^{0.5} \cdot C_M - k_2 \cdot C_X \cdot C_H^{0.5}$$

$$\frac{d}{d\tau}C_M = -k_1 \cdot C_H^{0.5} \cdot C_M$$

$$\frac{d}{d\tau}C_X = k_1 \cdot C_H^{0.5} \cdot C_M - k_2 \cdot C_X \cdot C_H^{0.5}$$

where τ is the space time (residence time), k_1 is the reaction constant of reaction 1, k_2 is the reaction constant of reaction 2, C_H, C_M, and C_X are the concentrations of hydrogen, mesitylene, and m-xylene at a specified τ in the reactor, respectively. The concentrations of hydrogen and mesitylene at the entrance are 0.021 and 0.0105 lbmol/ft³, respectively.

Data: $k_1 = 55.2$ (ft³/lbmol)$^{0.5}$hour^{-1} $k_2 = 30.2$ (ft³/lbmol)$^{0.5}$hour^{-1}

a. Plot the concentration of hydrogen, mesitylene, m-xylene as a function of τ (from 0 to 0.5 hr).
b. Determine the optimum space time of the plug flow reactor that gives maximum product.

Solution:

$$k_1 := 55.2 \quad \left(\frac{ft^3}{lbmol}\right)^{0.5} \cdot hr^{-1} \qquad k_2 := 30.2 \quad \left(\frac{ft^3}{lbmol}\right)^{0.5} \cdot hr^{-1}$$

Using *odesolve* function

Given

$$\frac{d}{d\tau}C_H(\tau) = -k_1 \cdot C_H(\tau)^{0.5} \cdot C_M(\tau) - k_2 \cdot C_X(\tau) \cdot C_H(\tau)^{0.5} \qquad C_H(0) = 0.021$$

$$\frac{d}{d\tau}C_M(\tau) = -k_1 \cdot C_H(\tau)^{0.5} \cdot C_M(\tau) \qquad\qquad\qquad C_M(0) = 0.0105$$

$$\frac{d}{d\tau}C_X(\tau) = k_1 \cdot C_H(\tau)^{0.5} \cdot C_M(\tau) - k_2 \cdot C_X(\tau) \cdot C_H(\tau)^{0.5} \qquad C_X(0) = 0$$

$$\begin{pmatrix} C_H \\ C_M \\ C_X \end{pmatrix} := \text{Odesolve} \left[\begin{pmatrix} C_H \\ C_M \\ C_X \end{pmatrix}, \tau, 0.5 \right]$$

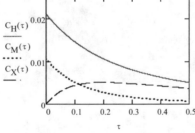

Initial guess: $\tau1 := 0.2$

$\tau1 := \text{Maximize}(C_X, \tau1)$

$\tau1 = 0.198$

The optimum τ is 0.198 hr

3. A small spherical organism with a radius (R) of 0.01 cm will not reproduce if an organ located at the center of the organism is exposed to a concentration of oxygen below 1.8 10^{-4} mol/cm³. The diffusivity of oxygen through the organism (D_{AB}) is 10^{-6} cm²/s. The respiration rate of this organism is a first order reaction:

$$-r_A = k \cdot C_A$$

where k is the rate constant (= 0.04 s^{-1}) and C_A is the oxygen concentration (mol/cm³).

a. If the oxygen concentration on the surface of the organism (C_R) is 2.5 10^{-4} mol/cm³, will this condition permit reproduction?
b. Determine the critical oxygen concentration on the surface to support the organism reproduction.

The following differential equation can be developed from mol balance of oxygen:

$$\frac{1}{r^2} \cdot \left[\frac{d}{dr} \left[r^2 \cdot \left(\frac{d}{dr}C_A \right) \right] \right] - \frac{k}{D_{AB}} \cdot C_A = 0$$

with boundary conditions: at r = 0: $\frac{d}{dr}C_A = 0$ at r = R: $C_A = C_R$

Solution:

$$D_{AB} := 10^{-6} \qquad k := 0.04 \qquad C_R := 2.5 \cdot 10^{-4} \qquad R := 10^{-2}$$

Using *odesolve* function

Given

$$\frac{d^2}{dr^2}C(r) + \frac{2}{r}\left(\frac{d}{dr}C(r)\right) - \frac{k}{D_{AB}} \cdot C(r) = 0$$

$$C\left(10^{-8}\right) = 0$$

$$C(R) = C_R$$

$$C\left(C_R\right) := \text{Odesolve}\,(r, R)$$

$$r := 10^{-8}, 0.0001.. R$$

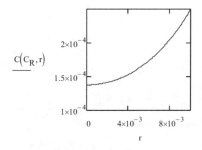

$$C\left(C_R, r\right)$$

a. $R0 := 10^{-8}$

$C\left(C_R, R0\right) = 1.379 \times 10^{-4} \ < 1.8\ 10^{-4}$

Thus, this condition does not permit reproduction.

b. $C_R := 3 \cdot 10^{-4}$

$C_R := \text{root}\left(\,C\left(C_R, R0\right) - 1.8\ 10^{-4}, C_R\right)$

$C_R = 3.264 \times 10^{-4} \quad \text{mol/cm}^3$

4. A straight fin of uniform rectangular cross section (0.5 mm x 100 mm) is attached to a base surface of temperature 110°C (T_b). The surface of the fin is exposed to a cooling fluid at 20°C (T_f) with a convection heat transfer coefficient (h) of 15 W/(m^2.K). The conductivity (k) of the fin material is 400 W/(m.K). If the length of the fin (L) is 5 cm, determine:
a. the temperature profile T(x) along the length of the fin
b. the heat transferred (q) by the fin

From energy balance, we obtain the following differential equation:

$$\frac{d^2}{dx^2}T - \frac{h \cdot P}{k \cdot A_c} \cdot (T - T_f) = 0$$

P is the fin perimeter, A_c is the fin cross-sectional area, and x is the distance measured from the base.

with boundary conditions:

The temperature at the base of the fin (x = 0): $\qquad T(0) = T_b$

The rate at which energy is transferred to the fluid by convection from the tip (x = L) must equal the rate at which energy reaches the tip by conduction through the fin:

$$h \cdot A_c \cdot \left(T(L) - T_f\right) = -k \cdot A_c \cdot \frac{dT}{dx}\bigg|_{x=L}$$

The heat transferred by the fin can be calculated once the temperature profile is known:

$$q = -k \cdot A_c \cdot \frac{dT}{dx}\bigg|_{x=0}$$

Solution: This is a second order ODE with boundary value problem

Define: $\quad \theta = T - T_f \qquad m^2 = \frac{h \cdot P}{k \cdot A_c}$

$$\frac{d^2}{dx^2}\theta - m^2 \cdot \theta = 0 \qquad \theta(0) = T_b - T_f \qquad h \cdot \theta(L) + k \cdot \theta'(L) = 0$$

Odesolve function cannot handle the second boundary condition.

$$h := 15 \qquad A_c := 50 \cdot 10^{-6} \qquad k := 400 \qquad P := 201 \cdot 10^{-3} \qquad m2 := \frac{h \cdot P}{k \cdot A_c}$$

$$T_b := 110 + 273.15 \qquad T_f := 20 + 273.15 \qquad m2 = 150.75$$

$$D(x,y) := \begin{pmatrix} y_1 \\ m2\,y_0 \end{pmatrix} \qquad vig_0 := 1 \qquad IV(a, vig) := \begin{pmatrix} T_b - T_f \\ vig_0 \end{pmatrix}$$

$$EV(b,y) := h \cdot y_0 + k \cdot y_1$$

$$s := sbval(vig, 0, 0.05, D, IV, EV)$$

$$n := 100 \qquad vyi := \begin{pmatrix} T_b - T_f \\ s_0 \end{pmatrix} \qquad u(x) := rkfixed(vyi, 0, x, n, D)$$

a.   Temperature profile

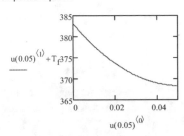

$$u(0.05)^{\langle 1 \rangle} + T_f$$

b.   The heat transfer by the fin:

$$q := -k \cdot A_c \cdot \left(u(0.05)^{\langle 2 \rangle} \right)_0$$

$$q = 12.133 \qquad \text{Watt}$$

8.2 Partial Differential Equation (PDE)

A partial differential equation is a differential equation that has more than one independent variable. The independent variables involved may be several coordinates in space, or time and one or several coordinates in space. A PDE is called an n-th order PDE if the highest derivative is of order n and is called linear if it is linear in the dependent variable and all of its partial derivatives.

The general linear second order PDE with two independent variables can be expressed as

$$a\frac{\partial^2 f}{\partial x^2} + 2b\frac{\partial^2 f}{\partial x \partial y} + c\frac{\partial^2 f}{\partial y^2} = d$$

If we define $\Delta = b^2 - 4ac$, reminiscent of the discriminant for a quadratic polynomial, then a PDE is classified as

elliptic   if $\Delta < 0$

parabolic   if $\Delta = 0$

hyperbolic   if $\Delta > 0$

In chemical engineering, we usually encounter second order parabolic and elliptic PDEs, such as

Fick's second law of diffusion: $\dfrac{\partial C}{\partial t} = D_{AB}\dfrac{\partial^2 C}{\partial z^2}$ parabolic

Laplace's equation for heat conduction: $\dfrac{\partial^2 T}{\partial x^2} + \dfrac{\partial^2 T}{\partial y^2} = 0$ elliptic

8.2.1 Parabolic PDE

To solve a parabolic PDE with Dirichlet $[f(x_0,t) = g(t)]$ and/or Neumann $[\partial f/\partial x(x_0,t) = h(t)]$ boundary conditions, *pdesolve* function can be directly used. Of course, for parabolic PDE, we must have two boundary conditions and one initial condition.

Since *pdesolve* function uses a numerical method, the solution obtained is an approximation. The values of the dependent variable (i.e., the solution) are saved at discrete points of the independent variables. If the value of the dependent variable is needed at independent variables other than the discrete points, the solution will be interpolated from a matrix of solution points. Procedure 8.12 shows the use of this function.

Procedure 8.12: *pdesolve*

Suppose we want to solve the following parabolic PDE for $0 \leq x \leq 5$ and $0 \leq t \leq 20$:

$$\frac{\partial}{\partial t}f = 0.2 \cdot \frac{\partial^2}{\partial x^2}f \quad \text{at } t = 0, \text{ all } x \quad f = 6 \quad \text{at } x = 0, \text{ all } t \quad \frac{\partial}{\partial x}f = 0$$

$$\text{at } x = 5, \text{ all } t \quad f = 5$$

1. Set up the Given block containing the PDE and its initial and boundary conditions:

Given

$$f_t(x,t) = 0.2 \cdot f_{xx}(x,t)$$

IC: $f(x,0) = 6$

BCs: $f_x(0,t) = 0$ Neumann

$f(5,t) = 5$ Dirichlet

Note: 1. The partial derivatives are typed using subscript notation.
2. The order of the arguments of functions must be (x,t), not (t,x)
3. It is a good practice to type the first derivative f_t on the left side and the others on the right side.

2. Solve the PDE using *pdesolve* function:

$$f := \text{Pdesolve}\left[f, x, \begin{pmatrix} 0 \\ 5 \end{pmatrix}, t, \begin{pmatrix} 0 \\ 20 \end{pmatrix}\right]$$

pdesolve function can be typed any way we want, such as *Pdesolve* or *PDEsolve*.

If the number of points of each independent variable needs to be set, say to 100, the syntax will be as follows:

$$\text{Pdesolve}\left[f, x, \begin{pmatrix} 0 \\ 5 \end{pmatrix}, t, \begin{pmatrix} 0 \\ 20 \end{pmatrix}, 100, 100\right]$$

The sixth argument is the number of points of x and the last argument is the number of points of t.

3. The solution can be plotted or used to calculate the dependent variable at certain independent variables:

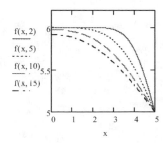

For example at x = 2 and t = 5:

$$f(2,5) = 5.966$$

Like *odesolve* function, *pdesolve* function can solve parametric problems when we have only one PDE. Parametric parabolic problems can also be solved using *numol* function by including the parameters in the arguments of the derivative function (similar to the derivative function in *rkfixed* function). However, *numol* function does not allow mixed boundary conditions; it solves problems with Dirichlet boundary conditions only or Neumann boundary conditions only. Procedure 8.13 describes the use of *numol* function.

Procedure 8.13: *numol*

Suppose we want to solve the following parabolic PDE for $0 \le x \le 5$ and $0 \le t \le 20$:

$$\frac{\partial}{\partial t} f = 0.2 \cdot \frac{\partial^2}{\partial x^2} f \qquad \text{at } t = 0, \text{ all } x \qquad f = 6$$

$$\text{at } x = 0, \text{ all } t \qquad f = 0$$

$$\text{at } x = 5, \text{ all } t \qquad f = 5$$

Usually t is the time variable and x is the variable for spatial coordinate.

1. Set up variables containing the starting and end values of the spatial coordinate and time:

$$xa := 0 \qquad xb := 5 \qquad ta := 0 \qquad tb := 20$$

2. Set up a derivative function containing the expression of f_t from the PDE:

$$D(x, t, f, f_x, f_{xx}) := 0.2 \cdot f_{xx} \qquad \text{Type \& order of the function arguments must be as shown.}$$

3. Set up functions containing initial condition and boundary conditions:

$$IC(x) := 6 \qquad \text{This function must have an argument representing the spatial coordinate.}$$

$$BC(t) := (\,0 \quad 5 \quad "D"\,) \qquad \text{This matrix function must have an argument representing the time variable.}$$

The first element of the matrix is the boundary condition at the starting point and the second is the boundary condition at the end point. The third element is the type of the boundary conditions, "D" is for Dirichlet and "N" is for Neumann.

3. Solve the PDE using *numol* function:

$$nx := 101 \qquad \text{The number of points for the spatial coordinate (\underline{not} the number of intervals)}$$

$$nt := 101 \qquad \text{The number of points for the time variable (\underline{not} the number of intervals)}$$

$$u := numol\left[\begin{pmatrix} xa \\ xb \end{pmatrix}, nx, \begin{pmatrix} ta \\ tb \end{pmatrix}, nt, 1, 0, D, IC, BC\right]$$

The first argument is a vector containing the starting and end values of x, the third is a vector containing the initial and end values of t, the fifth is the number of PDEs, and the sixth is the number of algebraic constraints (no algebraic constraint in this example)

u =	0	1	2	3	4
0	0	0	0	0	0
1	6	0.842	0.597	0.488	0.423
2	6	1.658	1.184	0.97	0.842

The values of the dependent variables as a function of the spatial coordinate for a certain time are given in a certain column of this matrix.

To plot the function, it would be better to use cubic spline and generate a fitting curve:

$$\Delta x := \frac{xb - xa}{nx - 1} \qquad \Delta t := \frac{tb - ta}{nt - 1} \qquad i := 0 .. nx - 1 \qquad j := 0 .. nt - 1$$

$$x_i := i \cdot \Delta x \qquad t_j := j \cdot \Delta t$$

$$Mxt := augment(x, t) \qquad vs := cspline(Mxt, u) \qquad f(x, t) := interp\left[vs, Mxt, u, \begin{pmatrix} x \\ t \end{pmatrix}\right]$$

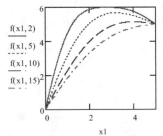

$f(x1, 2)$

$f(x1, 5)$

$f(x1, 10)$

$f(x1, 15)$

Plot of the function f at t = 2, 5, 10, and 15.

x1

8.2.2 System of Parabolic PDEs

Similar to *odesolve* function, *pdesolve* function can also be used to solve a system of parabolic PDEs, as shown in Procedure 18.14.

Procedure 8.14: *pdesolve for a system of parabolic PDEs*

Suppose we want to solve the following system of parabolic PDEs for $0 \leq r \leq 1$ and $0 \leq z \leq 1$:

$$\frac{\partial}{\partial z}f = 0.2\left[\frac{\partial^2}{\partial r^2}f + \frac{1}{r}\cdot\left(\frac{\partial}{\partial r}f\right)\right] - 200\cdot f\cdot\exp\left(\frac{-10}{g}\right)$$

at z = 0, all r $f = 1$ $g = 1$

$$\frac{\partial}{\partial z}g = 0.8\left[\frac{\partial^2}{\partial r^2}g + \frac{1}{r}\cdot\left(\frac{\partial}{\partial r}g\right)\right] + 2\cdot f\cdot\exp\left(\frac{-10}{g}\right)$$

at r = 0, all z $\frac{\partial}{\partial r}f = 0$ $\frac{\partial}{\partial r}g = 0$

at r = 1, all z $\frac{\partial}{\partial r}f = 0$ $-\left(\frac{\partial}{\partial r}g\right) = 2\cdot(g - 3)$

1. Set up the Given block containing the PDEs and their boundary conditions:

Given

$$f_z(r,z) = 0.2\cdot\left(f_{rr}(r,z) + \frac{1}{r}\cdot f_r(r,z)\right) - 200\cdot f(r,z)\cdot\exp\left(\frac{-10}{g(r,z)}\right)$$ $f(r,0) = 1$

$$g_z(r,z) = 0.8\cdot\left(g_{rr}(r,z) + \frac{1}{r}\cdot g_r(r,z)\right) + 2\cdot f(r,z)\cdot\exp\left(\frac{-10}{g(r,z)}\right)$$ $g(r,0) = 1$

$$f_r\left(10^{-12},z\right) = 0 \qquad g_r\left(10^{-12},z\right) = 0$$ $f_r(1,z) = 0$

$$-g_r(1,z) = 2\cdot(g(1,z) - 3)$$

2. Solve the PDEs using *pdesolve* function:

$$\begin{pmatrix} f \\ g \end{pmatrix} := \text{Pdesolve}\left[\begin{pmatrix} f \\ g \end{pmatrix}, r, \begin{pmatrix} 10^{-12} \\ 1 \end{pmatrix}, z, \begin{pmatrix} 0 \\ 1 \end{pmatrix}\right]$$

The third and fifth arguments are the ranges for r and z, respectively.

3. The solution can be plotted or used to calculate the dependent variable at certain independent variables:

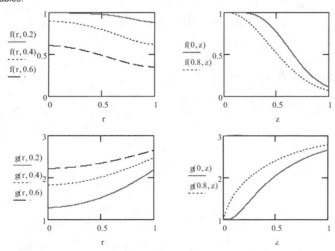

In Procedure 8.14, the numerical method used in *pdesolve* function has been changed from **Recursive 5-Point Differences** to **Polynomial**. The change of numerical method used is usually needed if we see some unphysical results. These methods can be accessed by right clicking on *pdesolve* and choose the desired method from the pop-up menu.

8.2.3 Elliptic PDE

To solve an elliptic PDE, Mathcad does not have a function that can be used directly to solve the PDE. We have to apply finite difference to approximate the derivatives and then use *relax* function. Procedure 8.15 shows the use of this function.

Procedure 8.15: *relax*

Suppose we want to solve the following PDE:

$$\frac{d^2}{dx^2}f + \frac{d^2}{dy^2}f = 0$$

with the boundary conditions: $f(0,y) = 2$ $f(x,0) = 5$ $f(L,y) = 0$ $f(x,L) = 10$

Using finite difference: $\dfrac{d^2}{dx^2}f = \dfrac{f_{i+1,j} - 2 \cdot f_{i,j} + f_{i-1,j}}{\Delta x^2}$ $\dfrac{d^2}{dy^2}f = \dfrac{f_{i,j+1} - 2 \cdot f_{i,j} + f_{i,j-1}}{\Delta y^2}$

we can transform the PDE into finite-difference equations:

$$f_{i+1,j} - 2 \cdot f_{i,j} + f_{i-1,j} + f_{i,j+1} - 2 \cdot f_{i,j} + f_{i,j-1} = 0 \qquad \text{(taking } \Delta x = \Delta y)$$

or

$$f_{i+1,j} + f_{i-1,j} + f_{i,j+1} + f_{i,j-1} - 4 \cdot f_{i,j} = 0$$

Δx is not necessarily equal to Δy

In Mathcad, the finite-difference equation should be written in the following form:

$$a \cdot f_{i+1,j} + b \cdot f_{i-1,j} + c \cdot f_{i,j+1} + d \cdot f_{i,j-1} + e \cdot f_{i,j} = s_{i,j}$$

1. Define the number of grids:

$\qquad\qquad N := 30 \qquad i := 0..N \qquad j := 0..N$

In using *relax* function, the number of grids in any spatial coordinate must be the same.

2. Define variables a through e for our case:

$\qquad\qquad a_{i,j} := 1 \quad b_{i,j} := 1 \quad c_{i,j} := 1 \quad d_{i,j} := 1 \quad e_{i,j} := -4$

3. Define the values of $s_{i,j}$:

$\qquad\qquad s_{i,j} := 0$

4. Set up the boundary conditions:

$\qquad\qquad f_{0,j} := 2 \qquad f_{i,0} := 5 \qquad f_{N,j} := 0 \qquad f_{i,N} := 10$

5. Define Jacobi spectral radius (r), which is a parameter that controls the convergence of the algorithm:

$\qquad\qquad r := 1 - \dfrac{2 \cdot \pi}{N}$

If this value gives the error message "too many iterations", try to reduce the value of r.

The value of r is usually between 0 and 1.

6. Solve the finite difference equation using *relax* function:

$\qquad\qquad F := \text{relax}(a,b,c,d,e,s,f,r)$

F is an (N+1) by (N+1) matrix

$F_{i,j}$ represents the value of f at

$\qquad x = (L/N)i$
$\qquad y = (L/N)j$

	0	1	2	3
0	5	2	2	2
1	5	3.504	2.915	2.64
2	5	4.1	3.515	3.153
3	5	4.382	3.891	...

$F = $ (preceding the table)

7. The results can be plotted in a 3-D graph (surface plot) or a 2-D graph (contour plot):

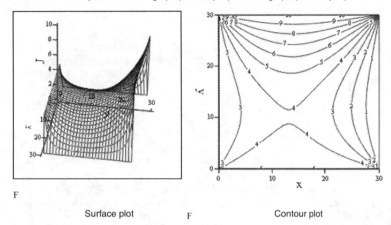

F

Surface plot   F   Contour plot

Example set 8.2

1. Initially, the concentration of A in a stagnant medium B of 6 cm thick is 0.1 mol/cm^3. A light is then incident on the system so that a photochemical reaction occurs. The destruction rate of A is given by:

$$N_A = -k \cdot C_A \text{ mol/(cm}^3\text{.s)}$$

where k is the reaction constant (=0.03 s^{-1}) and C_A is the concentration of A. One side of the medium is in contact with a pure gas A so that the concentration of A on this medium surface can be kept constant at 0.1 mol/cm^3. The other side of the medium is in contact with an impermeable wall. If the diffusivity of A in medium B (D_{AB}) is 0.015 cm^2/s, plot the concentration profile of A in the medium after 2, 5, and 10 s. What is the concentration of A at x = 0.25 cm after 8 s?

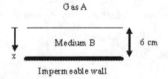

Gas A

Medium B   6 cm

x

Impermeable wall

From the mole balance, the following PDE is obtained:

$$\frac{d}{dt}C_A = D_{AB} \cdot \frac{d^2}{dx^2}C_A - k \cdot C_A$$

Initial condition: at t = 0, all x $C_A = 0.1$

Boundary conditions: at x = 0, all t $C_A = 0.1$

    at x = 6, all t $\frac{d}{dx}C_A = 0$ (impermeable wall)

where t is time in second and x is the distance from the gas-medium interface in cm.

<u>Solution:</u> $D_{AB} := 0.015$ $k := 0.03$

Given

$$C_t(x,t) = D_{AB} \cdot C_{xx}(x,t) - k \cdot C(x,t)$$

$$C(x,0) = 0.1$$

$$C(0,t) = 0.1 \qquad C_x(6,t) = 0$$

$$C := \text{Pdesolve}\left[C, x, \begin{pmatrix} 0 \\ 6 \end{pmatrix}, t, \begin{pmatrix} 0 \\ 10 \end{pmatrix}\right]$$

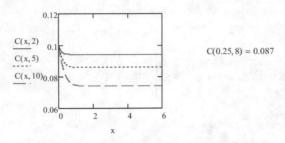

$C(0.25, 8) = 0.087$

2. The temperature of three surfaces of a long bar, 10 cm by 10 cm on a side, are maintained at 30ºC while the remaining surface is maintained at 20ºC by exposing to a colder air with a very high convective heat transfer coefficient At the center of the bar, an electric wire produces heat at a rate of 4 10⁶ W/m³ (q). Plot the two-dimensional temperature distribution in the bar at steady state if the conductivity (k) of the column material is 1 W/m.K.

From the energy balance, the following partial differential equation is obtained:

$$\frac{d^2}{dx^2}T + \frac{d^2}{dy^2}T + \frac{q}{k} = 0$$

with the boundary conditions: $T(0,y) = 30$ $T(0.1,y) = 30$

$$T(x,0) = 20 \qquad T(x,0.1) = 30$$

where T is the temperature and (x,y) is the coordinate.

Solution:

Using finite difference: $\dfrac{d^2}{dx^2}T = \dfrac{T_{i+1,j} - 2 \cdot T_{i,j} + T_{i-1,j}}{\Delta x^2}$ $\dfrac{d^2}{dy^2}T = \dfrac{T_{i,j+1} - 2 \cdot T_{i,j} + T_{i,j-1}}{\Delta y^2}$

we can transform the PDE into finite-difference equations:

$$T_{i+1,j} + T_{i-1,j} + T_{i,j+1} + T_{i,j-1} - 4 \cdot T_{i,j} = \frac{-q_{i,j} \cdot \Delta x^2}{k} \qquad \text{(taking } \Delta x = \Delta y)$$

$$N := 30 \qquad i := 0..N \qquad j := 0..N$$

By comparing the above finite-difference equation with the general equation in Mathcad:

$$a \cdot f_{i+1,j} + b \cdot f_{i-1,j} + c \cdot f_{i,j+1} + d \cdot f_{i,j-1} + e \cdot f_{i,j} = s_{i,j}$$

we get $a_{i,j} := 1 \quad b_{i,j} := 1 \quad c_{i,j} := 1 \quad d_{i,j} := 1 \quad e_{i,j} := -4$

Heat is produced at the center of the bar:

$$\Delta x := \frac{0.1}{N} \qquad q := 4 \cdot 10^6 \qquad k := 1 \qquad s_{i,j} := 0 \qquad s_{\frac{N}{2}, \frac{N}{2}} := \frac{-q \cdot \Delta x^2}{k}$$

The boundary conditions:

$$T_{0,j} := 30 \qquad T_{i,0} := 20 \qquad T_{N,j} := 30 \qquad T_{i,N} := 30$$

Solve the finite difference equation using *relax* function:

$$r := 1 - \frac{2 \cdot \pi}{N}$$

$$F := \text{relax}(a, b, c, d, e, s, T, r)$$

$$F = \begin{array}{c|c|c|c|c|} & 0 & 1 & 2 & 3 \\ \hline 0 & 20 & 30 & 30 & 30 \\ \hline 1 & 20 & 25.066 & 27.109 & 28.104 \\ \hline 2 & 20 & 23.156 & 25.264 & 26.629 \\ \hline 3 & 20 & 22.292 & 24.164 & ... \\ \hline \end{array}$$

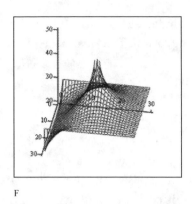

F

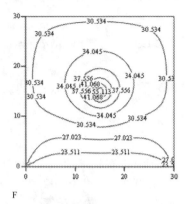

F

Problems

1. A liquid reaction will be carried out in a Continuous Stirred Tank Reactor (CSTR). Initially (at $t = 0$), the concentration of A in the tank is zero. Then, a feed stream containing 0.5 mol A/cm³ (C_{Ain}) enters the reactor, in which A dissociates into products with a first order reaction:

$$A \rightarrow products \qquad\qquad -r_A = kC_A$$

where $-r_A$ is the reaction rate [mol/(cm³.min)], k is the reaction constant [= 0.08 min⁻¹], and C_A is the concentration of A in the reactor [mol/cm³], which is the same as the concentration of A in the exit stream. If the residence time of the liquid inside the reactor (τ) is 10 minutes, plot the change in concentration of A in the exiting stream as a function of time ($0 \le t \le 5\tau$).

From mole balance, we can construct a differential equation for this startup process:

$$\frac{dC_A}{dt} + \frac{1+\tau k}{\tau}C_A = \frac{C_{Ain}}{\tau}$$

2. A tank initially contains 300 kg of salt solution (M at $t = 0$) and the mass fraction of salt in the solution is 8% (w at $t = 0$). Then, a stream containing 15% salt (w_i) enters the tank at a constant flow rate of 15 kg/h (F_i) and a stream at a constant flow rate of 8 kg/h (F_o) leaves the tank. Plot the total mass of the salt solution (M) vs. time (t) from $t = 0$ to 10 hours. Also create another graph for the mass fraction of salt (w) in the exit stream vs. time. What are the total mass of the solution in the tank and its salt concentration, which is equal to the concentra-

tion of salt in the exit stream, after 5 hours? If the tank is well mixed, the following differential equations can be derived from mass balance:

$$\frac{dM}{dt} = F_i - F_o$$

$$M\frac{dw}{dt} + w\frac{dM}{dt} = w_i F_i - wF_o$$

3. Four gas-phase reactions involving 6 substances occur simultaneously on a metal oxide supported catalyst [Fogler, *Elements of Chemical Reaction Engineering*, Prentice-Hall, Englewood Cliffs, 1992]. These reactions are carried out in a Plug Flow Reactor (PFR), in which their reaction rates are given as follows:

$$r_{1A} = -k_1 C_{T0}^3 \frac{F_A F_B^2}{F_T^3} \qquad r_{2A} = -k_2 C_{T0}^2 \frac{F_A F_B}{F_T^2}$$

$$r_{3B} = -k_3 C_{T0}^3 \frac{F_B F_C^2}{F_T^3} \qquad r_{4C} = -k_4 C_{T0}^{5/3} \frac{F_C F_A^{2/3}}{F_T^{5/3}}$$

In the above equations, C_{T0} is the total concentration at the entrance to the reactor, k_i is the reaction rate constant of reaction i, and F_T is the total molar flow rate given by

$$F_T = F_A + F_B + F_C + F_D + F_E + F_F$$

where F_A, F_B, F_C, F_D, F_E, F_F are the molar flow rates of components A, B, C, D, E, and F, respectively.

a. Determine the mol fraction of each substance in the mixture that comes out from the reactor if the reactor volume is 10 liter.
b. Plot the mol fraction profile of each substance in the reactor (the mol fraction as a function of volume from 0 to 10 liter).

Data: Molar flow rates at the entrance to the reactor:
$F_{A0} = F_{B0} = 10$ mol/min
$F_{C0} = F_{D0} = F_{E0} = F_{F0} = 0$

The total concentration at entrance to the reactor: $C_{T0} = 2$ mol/liter

Reaction rate constants: $k_1 = 5.0$ (liter/mol)²/min
$k_2 = 2.0$ liter/(mol·min)
$k_3 = 10.0$ (liter/mol)²/min
$k_4 = 5.0$ (liter/mol)^{2/3}/min

From mole balances, we can obtain:

$$\frac{dF_A}{dV} = r_{1A} + r_{2A} + \frac{2}{3}r_{4C} \qquad \frac{dF_B}{dV} = \frac{5}{4}r_{1A} + \frac{3}{4}r_{2A} + r_{3B}$$

$$\frac{dF_C}{dV} = -r_{1A} + 2r_{3B} + r_{4C} \qquad \frac{dF_D}{dV} = -\frac{3}{2}r_{1A} - \frac{3}{2}r_{2A} - r_{4C}$$

$$\frac{dF_E}{dV} = -\frac{1}{2}r_{2A} - \frac{5}{6}r_{4C} \qquad \frac{dF_F}{dV} = -2r_{3B}$$

The mol fraction of each species then can be calculated from:

$$X_i = \frac{F_i}{F_T}$$

4. In a membrane separation process, determine the membrane area needed to separate an air stream using a membrane 2.54×10^{-3} cm thick (t) with an oxygen permeability (P_A) of 500×10^{-10} cm^3 (STP)·cm/(s·cm^2·cmHg) and a separation factor (α), i.e., relative permeability of oxygen to that of nitrogen, of 10. The feed rate (L_f) is 1×10^6 cm^3 (STP)/s and the feed composition of oxygen (x_f) is 0.209. The pressure on the feed side (p_h) is 190 cmHg and on the permeate side (p_l) it is 19 cmHg. The desired composition of oxygen in the reject stream leaving the separator (x_o) is 0.119.

Using a cross-flow model, the following system of ODEs derived from material balance can be obtained:

$$\frac{dA_m}{dx} = \frac{y}{\dfrac{P_A'}{t}(p_h x - p_l y)} \frac{L_f - V}{x - y}$$

$$\frac{dV}{dx} = \frac{L_f - V}{x - y}$$

where dA_m is a differential membrane area, dV is the total gas flow rate permeating through the area dA_m, x is the composition of oxygen in the reject stream at a point along the stream path, and y is the corresponding composition of oxygen in the permeate side, which can be related to x by considering the mass-transfer rate through the membrane:

$$y = \frac{-\left[(\alpha-1)\left(x + \dfrac{p_l}{p_h}\right) + 1\right] + \sqrt{\left[(\alpha-1)\left(x + \dfrac{p_l}{p_h}\right) + 1\right]^2 - 4x\alpha\dfrac{p_l}{p_h}(\alpha-1)}}{2\dfrac{p_l}{p_h}(1-\alpha)}$$

The initial conditions for the ODEs are:

$$A_m(x = x_f) = 0 \text{ and } V(x = x_f) = 0$$

Thus, to calculate the total membrane area needed, the system of ODEs above must be solved simultaneously from $x = x_f$ to $x = x_o$.

5. The liquid level of a tank is controlled at a desired value when the inlet flow rate undergoes step change. As shown in the figure below, a feedback control system is used. This control system measures the liquid level and compares it with the desired steady-state value. If the level is higher than the desired value, it increases the effluent flow rate by opening the control valve, while it closes the valve when the level is lower than the desired value.

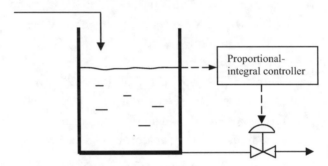

If the controller is a proportional-integral controller, the following second order ODE can be set up:

$$A\frac{d^2 \Delta h}{dt^2} + K_c \frac{d\Delta h}{dt} + \frac{K_c}{\tau_I}\Delta h = 0$$

where A is the cross-sectional area of the tank ($= 2 \text{ m}^2$), K_c is the proportional gain, τ_I is the integral time constant, t is time, and Δh is the deviation height of the liquid level in the tank, i.e., the difference between the actual height and the desired value.

For a certain step change of the inlet flow rate, the following initial conditions can be written:

$$\text{at t = 0: } \frac{d\Delta h}{dt} = 2 \text{ and } \Delta h = 0$$

a. Solve the ODE if K_c and τ_I are set to 1 m²/min and 0.1 min, respectively. Plot the Δh as a function of time to observe the dynamic behavior of the liquid level of the tank.

b. Using the worksheet in part a, vary the value of K_c (for example from 1 to 5) to learn about the effect of this parameter on the oscillatory behavior of the dynamic response of the liquid level due to a step change of the inlet flow rate. If the steady state value should be reached quickly, discuss whether or not a large value of K_c is better than a small one.

6. An important problem in chemical engineering is to predict the diffusion and reaction in a porous catalyst pellet. For first-order and irreversible reaction and negligible resistance of mass transfer across the boundary layer outside the pellets, mass and energy balances in a spherical domain give

$$\frac{1}{\left(r^*\right)^2}\frac{d}{dr^*}\left(\left(r^*\right)^2\frac{dC^*}{dr^*}\right) = \phi^2 C^* e^{\gamma-\frac{\gamma}{T^*}}$$

$$\frac{1}{\left(r^*\right)^2}\frac{d}{dr^*}\left(\left(r^*\right)^2\frac{dT^*}{dr^*}\right) = -\beta\phi^2 C^* e^{\gamma-\frac{\gamma}{T^*}}$$

BCs: $\left.\frac{dC^*}{dr^*}\right|_{r^*=0} = 0$ and $\left.C^*\right|_{r^*=1} = 1$

$\left.\frac{dT^*}{dr^*}\right|_{r^*=0} = 0$ and $\left.T^*\right|_{r^*=1} = 1$

where r^* = dimensionless radial distance from the pellet's center (= 0 at the center and 1 at the pellet radius)

C^* = dimensionless concentration (= 1 at the pellet radius)

T^* = dimensionless temperature (= 1 at the pellet radius)

ϕ^2 = Thiele modulus squared, which measures the relative importance of the reaction and diffusion phenomena. If the reaction is very fast, the Thiele modulus is large. If the diffusion is very fast, the Thiele modulus is small.

γ = dimensionless activation energy

β = dimensionless heat of reaction

Solve the ODE and plot the concentration and temperature profiles inside the catalyst pellet for $\beta = 0.3$, $\gamma = 18$, and $\phi = 0.5$.

Hint. For this complicated problem, we cannot use *Odesolve* function.

7. An 80 mm steel rod has a uniform temperature of 300 K. Suddenly, the temperature of one end is increased and kept constant at 400 K and the temperature of the other end is decreased and kept constant at 273 K. If the rod surface is assumed to be completely insulated, determine the temperature profile in the rod after 2, 5, and 8 seconds.

From the energy balance, this transient conduction can be represented by the following partial differential equation:

$$\frac{\partial T}{\partial t} = \alpha \frac{\partial^2 T}{\partial x^2}$$

In the above equation, T is the temperature, t is time, x is the axial distance, and α is the thermal diffusivity given by

$$\alpha = \frac{k}{\rho c_p}$$

where k is the thermal conductivity, ρ is the density, and c_p is the specific heat. For steel: $k = 63.9$ W/(m.K), $\rho = 7823$ kg/m³, and $c_p = 434$ J/(kg.K).

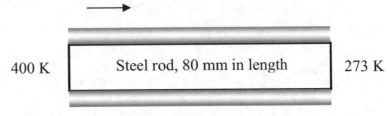

400 K   Steel rod, 80 mm in length   273 K

Insulation

8. A long bar, 0.3 cm by 0.2 cm on a side, contains dissolved hydrogen (H₂). The three surfaces of this bar are exposed to a fluid stream such that the concentrations of the dissolved hydrogen at these surfaces are zero. The remaining surface is contacted with hydrogen gas so that the concentration of the dissolved hydrogen at this surface can be maintained at 3 kmol/m³. Plot the concentration profile (the two-dimensional concentration distribution) of hydrogen across the cross section area of the bar and determine the concentration of hydrogen at the center.

From the material balance, the following partial differential equation can be obtained:

$$\frac{\partial^2 C}{\partial x^2} + \frac{\partial^2 C}{\partial y^2} = 0$$

with the boundary conditions:

$$C(x,0) = 3 \qquad C(0,y) = 0$$
$$C(0.3,y) = 0 \qquad C(x,0.2) = 0$$

where C is the hydrogen concentration and (x,y) is the coordinate.

9. For reactor start-up, it is often very important to analyze how temperature and concentration approach their steady-state values. In other

words, we need to know how temperature and concentration change with time during start-up. A significant overshoot in temperature may cause a reactant or product to degrade or the overshoot may be unacceptable for safe operation. Consider again the reacting system in the reactor described in Problem 3 of Chapter 4. For that system, it is known that an explosion can happen if the temperature of the reacting mixture is higher than 520 K. To analyze this issue, solve the system of differential equations shown in that problem ($0 \leq t \leq 0.3$ hr) and plot the Concentration-Temperature Phase Plane, i.e., $C_A(t)$ vs. $T(t)$. Initially (at $t = 0$), the concentration of A in the tank is 0.5 mol/cm³ and the temperature of the fluid in the tank is 300.15 K. Determine whether the start-up operation using these initial conditions is safe or not. If not, check whether or not decreasing the temperature of the cooling medium T_c to 269.15 K would make the operation safe.

10. A fixed-bed adsorber (silica gel) of 1.5 m deep (H_t) will be used to adsorb water from air at 21°C and 1 atm. The air containing 5×10^{-3} kg water/m³ (C_i) enters the column at a superficial velocity of 1000 m/hr (u). The amount of water in the exiting stream is not to exceed 0.25×10^{-3} kg water/m³ (C_o) and $K_c a$ for the mass transfer is estimated to be 7.2×10^4 hr^{-1}.

   a. Plot on the same graph the concentration profiles of water in the air inside the bed (C vs. z, where C is the water concentration in the air and z is the distance measured from the bed entrance) after $t = 5$, 10, and 50 hours.

   b. Determine how many hours are needed for the adsorption cycle (t_c), i.e., when $C(H_t, t_c) = C_o$ (when the water concentration in the air at the outlet reaches C_o).

   The vendor gives the following data for the silica gel used:

   Particle density (ρ_p) = 2000 kg/m³
   Bed porosity (ε) = 0.3

From material balance, we have:

Fluid side
$$\varepsilon \frac{\partial C}{\partial t} = -u \frac{\partial C}{\partial z} - K_c a\left(C - C^*\right)$$

Particle side
$$\rho_p(1 - \varepsilon)\frac{\partial q}{\partial t} = K_c a\left(C - C^*\right)$$

At $t = 0$: $C(z,0) = 0$, $q(z,0) = 0$
At $z = 0$: $C(0,t) = C_i$

where C is the equilibrium water concentration in kg/m³ on the fluid side at the solid-fluid interface. This equilibrium water concentration is related to the particle loading (q), i.e., the amount of water (in kg) adsorbed per kg silica gel, as follows

$$C^* = 0.031q$$

Note: For *Pdesolve* function, it is not clear what the default numbers of points for the independent variables are. For this problem, to obtain accurate results, set the number of points of each independent variable to 300 or more.

11. To study the dynamic behavior of two interacting tanks, the first tank with a cross-sectional area (A_1) of 2 m² is filled with water until the height of the liquid level is 1.5 m, while the second tank with a cross-sectional area (A_2) of 1 m² is filled with water until the height of the liquid level is 1.0 m. At $t = 0$, the valve below each tank is opened and the outlet flow rate of the first tank is F_1 and of the second tank is F_2. At the same time, 0.1 m³/min of water (F_i) enters the first tank. The outlet flow rates are related to the liquid levels in the tanks through the following equations:

$$F_1 = \frac{h_1 - h_2}{R_1} \qquad F_2 = \frac{h_2}{R_2}$$

where R_1 and R_2 are both 5 min/m². As expected, the height of the liquid level in each tank varies with time, so does the outlet flow rate of each tank. From material balance, the rate of change of the height of the liquid level in each tank can be derived:

$$A_1 \frac{dh_1}{dt} = F_i - F_1 \qquad A_2 \frac{dh_2}{dt} = F_1 - F_2$$

where h_1 is the height of the liquid level in the first tank and h_2 is the height of the liquid level in the second tank.

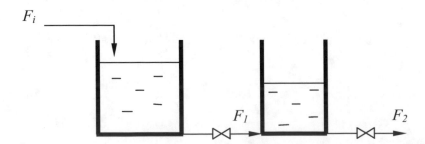

a. Plot the height of the liquid in each tank as a function of time, say from $t = 0$ to $t = 60$ min.

b. What is the maximum outlet flow rate of each tank?

Chapter 9
Miscellaneous

In this chapter, we discuss some other important features that are useful in many applications, i.e., data handling, data exchange with Excel, and Mathcad programming.

9.1 Data Handling

In some cases, we may need to read or write data from or to ASCII (text) files generated by other applications. The capability of Mathcad to transfer data from and to text files saves our time, especially when the number of data is large. There are several functions that can be used for this purpose, such as WRITEPRN, APPENDPRN, and READPRN.

WRITEPRN function is used to write data from a matrix variable in the active Mathcad worksheet to an ASCII (text) file. The data in the text file will be formatted in accordance with the matrix. For example, if the matrix in the Mathcad worksheet is a 50×40 matrix, the data in the text file will be arranged in 50 rows and 40 columns. APPENDPRN function is used to append data to the existing text data. READPRN function is used to read data from a text file and assign the data to a matrix variable in the worksheet. Procedure 9.1 shows how to use these file access functions.

9.2 Data Exchange with Excel

Data exchange with Excel worksheet can be done easily using *component* feature. This feature allows us not only to transfer data to/from the Excel worksheet, but also to do some calculations in Excel worksheet and transfer the results back to the active Mathcad worksheet. The Excel component thus performs a live data exchange. To those who are familiar with calculations in Excel worksheet, this Excel component might be handy. Procedure 9.2 demonstrates the data exchange with Excel.

Procedure 9.1: *WRITEPRN, APPENDPRN, and READPRN*

WRITEPRN

Suppose we want to write the following data to a text file in drive F:

$$A := \begin{pmatrix} 1 & 0 & 5 & -2 & 3 \\ 1 & 3 & -1 & 0 & 0 \\ 8 & 0 & 1 & 1 & 5 \\ 6 & -3 & 2 & 2 & 4 \\ 0 & 1 & 4 & 5 & -4 \end{pmatrix}$$ WRITEPRN("F:\exp.txt") := A

When we open the *exp.txt* in drive F using Notepad, we will see the following formatted data:

```
1    0    5    -2    3
1    3    -1    0     0
8    0    1     1     5
6   -3    2     2     4
0    1    4     5    -4
```

APPENDPRN

Suppose we want to add the following data to *exp.txt* in drive F that we have created:

$$B := \begin{pmatrix} 2 & 2 & 0 & -2 & 5 \\ 3 & 1 & 4 & 1 & 0 \\ -3 & 1 & 2 & 1 & -1 \end{pmatrix}$$ APPENDPRN("F:\exp.txt") := B

When we open the *exp.txt* in drive F using Notepad, we will see the following formatted data:

```
 1    0    5    -2    3
 1    3    -1    0     0
 8    0    1     1     5
 6   -3    2     2     4
 0    1    4     5    -4
 2    2    0    -2     5        The last three rows have been added
 3    1    4     1     0
-3    1    2     1    -1
```

READPRN

Suppose we want to read the data contained in *exp.txt* in drive F and assign the data to a vector variable C:

C := READPRN("F:\exp.txt")

$$C = \begin{pmatrix} 1 & 0 & 5 & -2 & 3 \\ 1 & 3 & -1 & 0 & 0 \\ 8 & 0 & 1 & 1 & 5 \\ 6 & -3 & 2 & 2 & 4 \\ 0 & 1 & 4 & 5 & -4 \\ 2 & 2 & 0 & -2 & 5 \\ 3 & 1 & 4 & 1 & 0 \\ -3 & 1 & 2 & 1 & -1 \end{pmatrix}$$

To extract a portion of the data from C, we can use *submatrix* function. For example if we want to obtain matrix A and B used previously:

A := submatrix(C, 0, 4, 0, 4)   B := submatrix(C, 5, 7, 0, 4)

The second and third arguments are row indices and the fourth and fifth are column indices.

$$A = \begin{pmatrix} 1 & 0 & 5 & -2 & 3 \\ 1 & 3 & -1 & 0 & 0 \\ 8 & 0 & 1 & 1 & 5 \\ 6 & -3 & 2 & 2 & 4 \\ 0 & 1 & 4 & 5 & -4 \end{pmatrix}$$ $$B = \begin{pmatrix} 2 & 2 & 0 & -2 & 5 \\ 3 & 1 & 4 & 1 & 0 \\ -3 & 1 & 2 & 1 & -1 \end{pmatrix}$$

Procedure 9.2: *Data Exchange with Excel using Excel component*

Transfering data from an Excel file

Suppose we have an Excel file (Book1.xls), the data of which will be transfered:

1. Insert an Excel component by clicking **Insert Component**, and selecting **Microsoft Excel** as the component to be inserted.

2. Click **Next** to open **Excel Setup Wizard** Window.

3. Choose **Create from file** and click **Browse** to find the Excel file. Choose the file and click **Open** then **Next**.

4. Define the number of input and output variables. Since we just want to transfer data from the Excel component, the number of input variables is zero. Suppose that we want to have 3 output variables, one contains the values in cells A1:A5, another contains the squares of the values in cells A1:A5, which will be put in cells B1:B5, and the other contains the sum of the squares, which will be put in cell B7. Type the cell range of each of the output variable in the given table and click **Finish**.

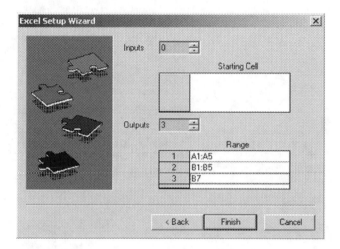

5. The Excel component is inserted, as shown below. Type in the variable names in the empty place holders, e.g., x, y, and z

$$\begin{pmatrix} x \\ y \\ z \end{pmatrix} := \begin{array}{|c|} \hline \\ \hline 2 \\ \hline 3 \\ \hline 4 \\ \hline 5 \\ \hline 6 \\ \hline \\ \hline \end{array}$$

6. At this stage, in this example, of course y is a zero vector, the elements of which are all zero, and z is also zero because the values of cells B1:B5 and B7 are still zero. To calculate the squares and sum of the squares in the Excel worksheet, double click the component and the Excel worksheet will be active. Then, any operation in the Excel worksheet can be performed. The results are directly assigned to x, y, and z.

$$\begin{pmatrix} x \\ y \\ z \end{pmatrix} :=$$

2	4
3	9
4	16
5	25
6	36
Sum	90

$$x = \begin{pmatrix} 2 \\ 3 \\ 4 \\ 5 \\ 6 \end{pmatrix} \qquad y = \begin{pmatrix} 4 \\ 9 \\ 16 \\ 25 \\ 36 \end{pmatrix} \qquad z = 90$$

Of couse, in fact the calculations of the squares and sum of the squares can be calculated in Mathcad once we have transfered the data and assigned them to variable x, but here we calculate those quantities in Excel component as an example.

Transfering data to and from an Excel file

Suppose we have a vector x, the element of which will be transfered to an Excel file. In the Excel file, for example, we calculate the squares and the sum of squares and transfer the results back to the Mathcad worksheet.

$$x := \begin{pmatrix} 2 \\ 3 \\ 4 \\ 1 \\ 2.5 \end{pmatrix}$$

We repeat the procedure for transfering data from an Excel file described previously, but now we choose **Create an empty Excel worksheet** instead of **Create from file** and the number of input variables is one. Note that the empty place holder below the component is for the name of the input variable.

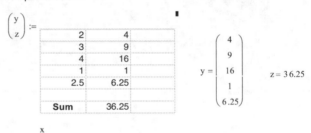

$$\begin{pmatrix} y \\ z \end{pmatrix} :=$$

2	4	
3	9	
4	16	
1	1	
2.5	6.25	
Sum	36.25	

x

$$y = \begin{pmatrix} 4 \\ 9 \\ 16 \\ 1 \\ 6.25 \end{pmatrix} \qquad z = 36.25$$

If the vector x is changed, it means the input is changed, then the output will automatically change. Thus, this component performs a live data exchange.

9.3 Introduction to Mathcad Programming

In Mathcad, sometimes we need to create a program for a certain calculation, especially when the calculation requires multi-step operation that involves conditionals and/or loops. A Mathcad program is in some sense similar to a subprogram in other programming languages, such as FORTRAN and C. Although programming in Mathcad is not as versatile as in FORTRAN and C, it offers a unique programming capability that can take the benefits of the built-in functions.

A Mathcad program can have one or more input variables that are passed into the program as arguments. Thus, a Mathcad program can be considered as a user-defined function with input variables as its arguments. Unlike a function in FORTRAN, which allows us to have only one output value, a Mathcad program can have more than one output values because a variable in Mathcad could serve as a vector or a matrix and thus the output values can be returned as vector or matrix elements.

9.3.1. Conditional (*if* Statement)

In a program containing conditionals, there are several choices of executions/operations and a decision to perform one of these operations should be made according to the input variables. Procedure 9.3 shows a program with *if* statement.

Procedure 9.3: *if* statement (logical if)

Suppose we want to define a function having this definition:
$$f(x) = 0 \text{ if } x < 2$$
$$= 1 \text{ if } 2 \le x \le 5$$
$$= 3 \text{ if } x > 5$$

1. Type the function name, the argument, and a colon
2. Click on the Programming Toolbar
3. Click **Add Line** and **if** icons
4. Type the value or statement to be executed on the left empty place holder and the conditional statement on the right empty place holder. Repeat 3-4 as needed

$$f(x) := \begin{vmatrix} 0 & \text{if } x < 2 \\ 1 & \text{if } 2 \le x \le 5 \\ 3 & \text{otherwise} \end{vmatrix}$$

$$f(1) = 0 \qquad f(4) = 1 \qquad f(6) = 3$$

5. Click *otherwise* icon as the last condition, which is executed only when all previous conditions are false.

Note: If there is an equal sign in the condition, the equal sign must be the Boolean equal sign.

It is important to remember that in all Mathcad programming, the output value returned is the value assigned in the last statement executed. In Procedure 9.3, when x = 1, the conditional statement in line 1 (x < 2) is evaluated first and since this condition is true, then the statement on the left side of this *if* statement is executed and 0 is assigned as the output value. The conditional statement in line 2 is then evaluated but since the condition is false, the statement on the left side is not executed. The statement in line 3 is not executed because one of the previous conditions is true. This corresponds to the *logical if* statement in FORTRAN, where only one statement needs to be executed for a certain condition.

In Procedure 9.3, without programming, in fact we could use *if* 'function' to define $f(x) := \text{if}(x < 2, 0, \text{if}(x \le 5, 1, 3))$. The first argument of this 'function' is the conditional statement (x < 2). The second argument will be executed if the conditional statement is true, otherwise the third argument will be executed. We need nested *if* 'function' because we have to test a second conditional statement (x ≤ 5) when the first conditional statement is false.

In many applications, instead of only one statement, several statements need to be executed for a certain condition. In this case, *if* 'function' cannot be used. We must use *if* statement in a program. Procedure 9.4 shows how to use *if* statement of this kind.

Procedure 9.4: *if* statement (if-then-otherwise)

Suppose we want to define a function having this definition:

if $x \geq 0$ $f(x) = a - 1$ if $a < 5$ if $x < 0$ $f(x) = -9$ if $a < -5$

$f(x) = a$ if $a \geq 5$ $f(x) = -1$ if $a \geq -5$

where $a = \sqrt{x}$ where $a = \sqrt{-x}$

$$
f(x) := \begin{vmatrix} \text{if } x \geq 0 \\ \quad \begin{vmatrix} a \leftarrow \sqrt{x} \\ a - 1 \text{ if } a < 5 \\ a \text{ otherwise} \end{vmatrix} \\ \text{otherwise} \\ \quad \begin{vmatrix} a \leftarrow \sqrt{-x} \\ -9 \text{ if } a < -5 \\ -1 \text{ otherwise} \end{vmatrix} \end{vmatrix}
$$

1. Type the function name, the argument, and a colon.
2. Click **Add Line** and **if** icons on the Programming Toolbar.
3. Type the condition on the right empty place holder.
4. To create several statements that need to be executed when the condition is true, click on the left empty place holder and click **Add Line**.

In a program, assignment/definition is done by using back arrow, which can be accessed from Programming Toolbar.

$f(3) = 0.732$ $f(-40) = -9$

The *if* statement in Procedure 9.4 can be called *if-(then)-otherwise* statement because it corresponds to *if-then-else* statement in FORTRAN. In Procedure 9.4, variable a is called a local variable, the value of which is undefined outside the program and should be defined inside the program. If a local variable is not defined inside a program and if its name has been defined somewhere outside the program (before the program), its value will be taken from outside the program, which could lead to unexpected error.

9.3.2. Loops (*for/while* Statement)

In a program containing loops, certain executions or operations are repeated a number of times before results are eventually returned. There are two kinds of loops in Mathcad: *for*-loop and *while*-loop. We use *for*-loop when the number of repetitions is known. The *while*-loop executes a block of statements repeatedly while a logical condition remains true. The *for*-loop is equivalent to *DO* loop and the *while*-loop is equivalent to *DO WHILE* loop in FORTRAN. Procedure 9.5 demonstrates *for* and *while* statements.

Procedure 9.5: *for/while* statement (Mathcad programming for loops)

Suppose we want to calculate a factorial:

$$
f(n) = \prod_{i=1}^{n} i
$$

In this case, we could directly use factorial operator $f(n) := n!$, but we will use a program here for learning the programming feature.

$$
f(n) := \begin{vmatrix} a \leftarrow 1 \\ \text{for } i \in 1..n \\ \quad a \leftarrow a \cdot i \\ a \leftarrow 1 \text{ if } n = 0 \end{vmatrix}
$$

1. Assignment/definition is done by using back arrow.
2. The loop is done by using a *for* statement, which is available in the Programming Toolbar.
4. i is a variable acting as the counter.
5. $a \leftarrow a \cdot i$ is executed for i =1 to n.
6. We need to include the *if* statement in the last line for n = 0. Keep in mind that the loop is still executed even though the counter is descending.

$f(3) = 6$ $f(0) = 1$

This operation can also be done using *while* statement:

$$f(n) := \begin{vmatrix} a \leftarrow 1 \\ i \leftarrow 1 \\ \text{while } i \leq n \\ \quad \begin{vmatrix} a \leftarrow a \cdot i \\ i \leftarrow i + 1 \end{vmatrix} \\ a \end{vmatrix}$$

$$f(3) = 6 \qquad f(0) = 1$$

1. The loop is done by using a *while* statement, which is available in the Programming Toolbar.
2. i is a variable acting as the test variable, which **must** be initialized before the *while* statement.
3. $a \leftarrow a \cdot i$ is executed repeatedly until $i > n$.
4. The test variable **must** be updated inside the loop, otherwise the calculations will never end.
5. We have to type the variable a in the last line, otherwise the last line is $i \leftarrow i + 1$ and the value of i will be returned as the value of f(n).

Example set 9.1

1. In a fermentation process, a CO_2-rich vapor containing a small amount of ethyl alcohol is evolved. The alcohol is then recovered by absorption with pure water in a sieve-tray tower. The molar flow rate of the pure water (L') is 144.9 kmol/h. The molar flow rate of CO_2 (V') in a gas mixture (CO_2/C_2H_5OH) entering the tower is estimated to be 176.4 kmol/h. The mol fraction (solute-free basis) of ethyl alcohol in this entering gas (Y_{N+1}) is 0.02. The desired mol fraction (solute-free basis) of ethyl alcohol in the gas stream exiting the tower (Y_1) is 0.0006.

At the condition of interest, the equilibrium curve that relates the mol fraction of ethyl alcohol in the gas phase leaving tray i (Y_i) to that in the liquid phase leaving the same tray (X_i) can be obtained from thermodynamics and is given by

$$Y_i = \frac{K \cdot X_i}{1 + (1 - K) \cdot X_i}$$

where K = 0.573. The operating line that relates the mol fraction of ethyl alcohol in the gas phase entering tray i (Y_{i+1}) to that in the liquid phase leaving the same tray (X_i) can be obtained from mole balance and is given by

$$Y_{i+1} = \frac{L'}{V'} \cdot X_i + \left(Y_1 - \frac{L'}{V'} \cdot X_0 \right) \qquad X_0 = 0 \quad \text{for pure water}$$

a. Determine the mol fraction (solute-free basis) of ethyl alcohol in the water stream leaving the tower (X_N) using the operating line.

b. Create a program to obtain the mol fraction of ethyl alcohol in the liquid phase leaving a tray (X_i) and that in the gas phase entering the tray (Y_{i+1}) for each tray.

c. Plot the stage-by-stage mol fractions of ethyl alcohol in Y vs. X diagram. Also plot the equilibrium and operating lines on the same graph.

The calculation can be done tray by tray. At first we know X_0 and Y_1. Then X_1 can be calculated from the equilibrium curve because we know Y_1. After X_1 is known, we can calculate Y_2 from the operating line. The procedure is repeated to obtain X_2, then Y_3, etc. until X_i exceeds X_N.

Solution: $V' := 176.4 \qquad Y_1 := 0.0006 \qquad YN_1 := 0.0204 \qquad X_0 := 0 \qquad$ (N1 to represent N+1)

$$K := 0.573 \qquad L' := 144.9$$

The equilibrium curve: $X_{eq}(Y) := \dfrac{Y}{K - (1 - K) \cdot Y}$

Since X_i will be obtained from Y_i, it is better to have X as a function of Y for the equilibrium curve.

The operating line: $Y_{op}(X) := \dfrac{L'}{V'} \cdot X + \left(Y_1 - \dfrac{L'}{V'} \cdot X_0 \right)$

a. The mole fraction of ethyl alcohol in the water stream leaving the tower:

$$XN := \frac{V'}{L'} \cdot \left[YN_1 - \left(Y_1 - \frac{L'}{V'} \cdot X_0 \right) \right] = 0.024$$

b. Create the program with three input variables, i.e., X0, Y1, and XN.

1. Start with i = 0
2. Define X_0 and Y_1 (the values of these variables must come from the input variables) and also save X_0 and Y_1 as output values. Since we have more than one output values, we must use a matrix variable to store all the output values.
3. Increase the value of i by 1.
4. From the known value of Y_i, obtain X_i from the equilibrium curve.

5. After X_i has been calculated, calculate Y_{i+1} from the operating line.

6. Save X_i and Y_{i+1} as output values and repeat steps 3 through 6 while $X_i < XN$

$$XY(X0, Y1, XN) := \begin{vmatrix} i \leftarrow 0 \\ X_0 \leftarrow X0 \\ Y_1 \leftarrow Y1 \\ A_{i,0} \leftarrow X_0 \\ A_{i,1} \leftarrow Y_1 \\ \text{while } X_i < XN \\ \quad \begin{vmatrix} i \leftarrow i + 1 \\ X_i \leftarrow X_{eq}(Y_i) \\ Y_{i+1} \leftarrow Y_{op}(X_i) \\ A_{i,0} \leftarrow X_i \\ A_{i,1} \leftarrow Y_{i+1} \end{vmatrix} \\ A \end{vmatrix}$$

Initialize index - step 1

Define X_0 and Y_1 and save the first pair of mole fractions in matrix A (X is in the first column and Y is in the second column) - step 2

Use *while* loop to calculate mole fractions in every tray (stop when $X_i \geq XN$) - steps 3 through 6

X_i is in equilibrum with Y_i

Y_{i+1} and X_i must be on the operating line

Save pair of mole fractions in matrix A (X is in the first column and Y is in the second column)

$M := XY(X0, Y1, XN)$ Matrix M contains pairs of mole fractions (X_i, Y_{i+1})

$$M^T = \begin{pmatrix} 0 & 0.001 & 0.0026 & 0.0047 & 0.0078 & 0.0123 & 0.0189 & 0.0285 \\ 0.0006 & 0.0015 & 0.0027 & 0.0045 & 0.007 & 0.0107 & 0.0161 & 0.024 \end{pmatrix}$$

c. Plot the stage-by-stage calculations in XY diagram:

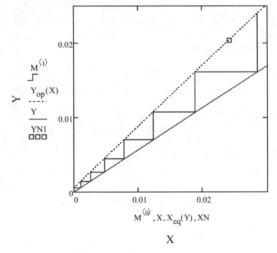

In this figure, the type of plot for M has been changed from *line* to *step*.

⌐ Stages
----- Operating line
——— Equilibrium line
□□□ Desired composition at bottom

2. Develop a function to calculate the molar volume of a pure fluid using Peng-Robinson equation of state. See Example 1 of Example set 3.1, Problems 1 and 2 of Chapter 3, and Problem 3 of Chapter 6.

The function for molar volume should be able to handle the following:

a. the calculation of molar volume in one phase region where only one positive real root is found (the other two roots could be all complex roots or negative real roots).

b. the calculation of molar volume in the region close to the saturated condition where three real roots are found.

The flow chart below helps us developing such a function:

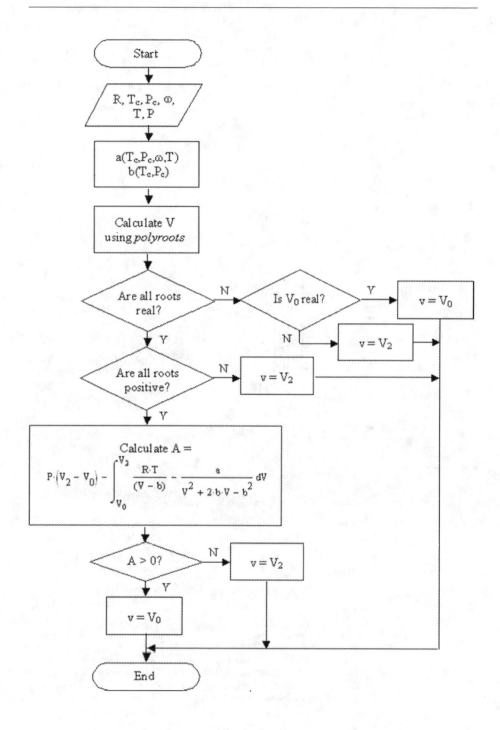

Solution:

$$v_{PR}(Tc, Pc, \omega, T, P) := \Bigg| R \leftarrow 83.14$$

$$a \leftarrow \frac{0.457235 R^2 \cdot Tc^2}{Pc} \cdot \left[1 + \left(0.37464 + 1.54226\omega - 0.26992\omega^2\right) \cdot \left[1 - \left(\frac{T}{Tc}\right)^{0.5}\right]\right]^2$$

$$b \leftarrow 0.077796 R \cdot \frac{Tc}{Pc}$$

$$c \leftarrow \begin{pmatrix} P \cdot b^3 + R \cdot T \cdot b^2 - a \cdot b \\ -3 \cdot P \cdot b^2 - 2 \cdot R \cdot T \cdot b + a \\ P \cdot b - R \cdot T \\ P \end{pmatrix}$$

$$V \leftarrow \text{polyroots (c)}$$

$$\text{if } Im(V) = \begin{pmatrix} 0 \\ 0 \\ 0 \end{pmatrix}$$

$$\Bigg| \text{if } V_0 > 0$$

$$A \leftarrow P \cdot (V_2 - V_0) - \int_{V_0}^{V_2} \frac{R \cdot T}{V - b} - \frac{a}{V^2 + 2 \cdot b \cdot V - b^2} \, dV$$

$$V_0 \text{ if } A > 0$$

$$V_2 \text{ otherwise}$$

$$V_2 \text{ otherwise}$$

$$\text{otherwise}$$

$$\Bigg| V_0 \text{ if } Im(V_0) = 0$$

$$V_2 \text{ otherwise}$$

Im function extracts the imaginary part of a complex number. If Im(x) returns zero, it means x is a real number.

Examples of molar volume calculations for carbon dioxide using this function:

$Tc := 304.2 \qquad Pc := 73.83 \qquad \omega := 0.224$

$v_{PR}(Tc, Pc, \omega, 283, 40) = 399.56$	There are three real roots, vapor phase
$v_{PR}(Tc, Pc, \omega, 283, 50) = 52.633$	There are three real roots, liquid phase
$v_{PR}(Tc, Pc, \omega, 283, 10) = 2196.735$	There is only one real root
$v_{PR}(Tc, Pc, \omega, 283, 100) = 47.972$	There is only one real root
$v_{PR}(Tc, Pc, \omega, 283, 2500) = 31.778$	There are three real roots, but only one positive real root

3. Two feed streams of benzene-toluene mixtures enters a distillation column, which operates at 1 atm, to produce a liquid distillate a liquid bottoms product of 95 mol% ($x_D = 0.95$) and 5 mol% ($x_W = 0.05$) benzene, respectively. The condenser used is a total condenser. The molar flow rate of the first feed (F_1), which is at its bubble-point temperature ($q_{F1} = 1$) and contains 75 mol% benzene ($z_{F1} = 0.75$), is 100 kgmol/hr. The second feed, which is an equimolar mixture ($z_{F2} = 0.5$), is preheated such that it enters the column with a molar liquid fraction of 0.6 (q_{F2}). If the molar flow rate of this second feed (F_2) is also 100 kgmol/hr, determine:

a. the molar flow rates of distillate (D) and bottoms (W)
b. the reflux ratio needed (R) if the column has 12 theoretical plates.

Equilibrium data for benzene/toluene at 1 atm can be represented by the following equation obtained from the fitting of the experimental data:

$$y_{eq}(x) = 7.806 \cdot 10^{-4} + 2.337 \cdot x - 3.402 x^2 + 7.481 \cdot x^3 - 13.766 x^4 + 12.761 \cdot x^5 - 4.412 \cdot x^6$$

The molar flow rates of distillate and bottoms can be determined by solving the following system of equations obtained from mole balance:

$$F_1 + F_2 = D + W$$

$$F_1 \cdot z_{F1} + F_2 \cdot z_{F2} = D \cdot x_D + W \cdot x_W$$

To obtain the reflux ratio for a given number of theoretical plates, the common method is usually to use a trial-and-error approach with graphical stage calculations, as shown in the xy diagram below:

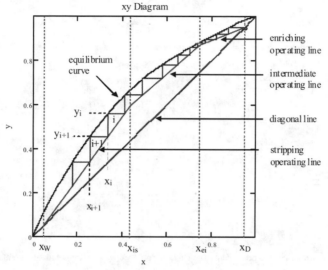

By assuming a certain value of R, the operating line for each section (enriching, intermediate, or stripping) is determined as follows:

Enriching $$y = \frac{R}{R + 1} \cdot x + \frac{x_D}{R + 1}$$

Intermediate $$y = \frac{L_i}{V_i} \cdot x + \frac{D \cdot x_D - F_1 \cdot z_{F1}}{V_i}$$

Stripping $$y = \frac{L_s}{V_s} \cdot x - \frac{W \cdot x_W}{V_s}$$

where the molar flow rates of liquid (L) and vapor (V) in each section are calculated from:

Enriching $\quad L_e = R \cdot D \qquad\qquad V_e = L_e + D$

Intermediate $\quad L_i = L_e + q_{F1} \cdot F_1 \qquad V_i = V_e - (1 - q_{F1}) \cdot F_1$

Stripping $\quad L_s = L_i + q_{F2} \cdot F_2 \qquad V_s = V_i - (1 - q_{F2}) \cdot F_2$

The subscripts e, i, and s refer to enriching, intermediate, and stripping section, respectively.

After all of the operating lines are defined, the points of intersection, i.e., x_{ei} and x_{is}, can be calculated and thus the operating line for the whole column is fixed (see diagram above). Then, stages can be drawn, stepping off from (x_0, y_1). For this problem, $x_0 = x_D$ and $y_1 = x_D$. This stage-by-stage drawing is similar to that in absorption (see Example 1 in Example set 9.1).

The calculation can be done tray by tray. At first we know x_0 and y_1. Then x_1 can be calculated from the equilibrium curve because the liquid leaving tray 1 is in equilibrium with the vapor leaving the same tray and we know y_1. After x_1 is known, we can calculate y_2 from the operating line in that section. The procedure is repeated to obtain x_2, then y_3, etc. until N trays.

As shown in the xy diagram above, the assumed value of R is still incorrect because x_{12} is not equal to x_W. Thus, different R should be used and the procedure is repeated until $x_{12} = x_W$. Of course, this is a tedious task. As we can see later, this can be solved easily by using a Mathcad program.

Solution: $F_1 := 100$ $z_{F1} := 0.75$ $q_{F1} := 1$ $x_D := 0.95$ $x_W := 0.05$

$F_2 := 100$ $z_{F2} := 0.5$ $q_{F2} := 0.6$ $N := 12$

a. the molar flow rates of distillate (D) and bottoms (W)

Initial guesses: $D := 100$ $W := 100$

Given

$$F_1 + F_2 = D + W$$

$$F_1 \cdot z_{F1} + F_2 \cdot z_{F2} = D \cdot x_D + W \cdot x_W$$

$$\begin{pmatrix} D \\ W \end{pmatrix} := Find(D, W) = \begin{pmatrix} 127.778 \\ 72.222 \end{pmatrix} \quad kgmol/hr$$

b. The operating line of each section in the column needs to be determined:

Molar flow rates of liquid and vapor in each section:

Enriching $L_e(R) := R \cdot D$ $V_e(R) := L_e(R) + D$

Intermediate $L_i(R) := L_e(R) + q_{F1} \cdot F_1$ $V_i(R) := V_e(R) - \left(1 - q_{F1}\right) \cdot F_1$

Stripping $L_s(R) := L_i(R) + q_{F2} \cdot F_2$ $V_s(R) := V_i(R) - \left(1 - q_{F2}\right) \cdot F_2$

Operating lines:

Enriching $y_{op1}(x, R) := \dfrac{R}{R+1} \cdot x + \dfrac{x_D}{R+1}$

Note that all of the flow rates and operating lines are functions of the reflux ratio R.

Intermediate $y_{op2}(x, R) := \dfrac{L_i(R)}{V_i(R)} \cdot x + \dfrac{D \cdot x_D - F_1 \cdot z_{F1}}{V_i(R)}$

Stripping $y_{op3}(x, R) := \dfrac{L_s(R)}{V_s(R)} \cdot x - \dfrac{W \cdot x_W}{V_s(R)}$

Determine the point of intersection between enriching and intermediate operating lines:

Initial guess: $x_{ei} := 0.9$

Given
$$y_{op1}\left(x_{ei}, R\right) = y_{op2}\left(x_{ei}, R\right) \qquad x_{ei}(R) := Find\left(x_{ei}\right)$$

Initial guess: $x_{is} := 0.5$

Given
$$y_{op2}\left(x_{is}, R\right) = y_{op3}\left(x_{is}, R\right) \qquad x_{is}(R) := Find\left(x_{is}\right)$$

The equilibrium curve:

$$y_{eq}(x) := 7.806 \cdot 10^{-4} + 2.337 x - 3.402 x^2 + 7.481 x^3 - 13.766 x^4 + 12.761 x^5 - 4.412 x^6$$

Create a program to calculate pairs of mole fractions of A in passing streams in the column:

$$xy\left(x_D, x_W, x_{ei}, x_{is}, N, R\right) :=
\begin{vmatrix}
A_{0,0} \leftarrow x_D \\
A_{0,1} \leftarrow x_D \\
x_0 \leftarrow x_D \\
y_1 \leftarrow x_D \\
\text{for } i \in 1 .. N \\
\quad \begin{vmatrix}
xi \leftarrow y_i \\
x_i \leftarrow root\left(y_{eq}(xi) - y_i, xi\right) \\
y_{i+1} \leftarrow y_{op1}\left(x_i, R\right) \quad \text{if } x_i > x_{ei} \\
y_{i+1} \leftarrow y_{op2}\left(x_i, R\right) \quad \text{if } x_{is} < x_i < x_{ei} \\
y_{i+1} \leftarrow y_{op3}\left(x_i, R\right) \quad \text{otherwise} \\
A_{i,0} \leftarrow x_i \\
A_{i,1} \leftarrow y_{i+1}
\end{vmatrix} \\
A
\end{vmatrix}$$

For example, if $\quad R := 0.7 \qquad M(R) := xy\big(x_D, x_W, x_{ei}(R), x_{is}(R), N, R\big)$

	0	1	2	3	4	5	6	7
0	0.95	0.8781	0.8203	0.775	0.74	0.705	0.6515	0.5778
1	0.95	0.9204	0.8966	0.8779	0.8589	0.8284	0.7817	...

$M(R)^T =$

Plot the stage-by-stage calculations in xy diagram: $\qquad y_{diag}(x) := x$

$$x_s := x_W, x_{is}(R) .. x_{is}(R) \qquad x_i := x_{is}(R), x_{ei}(R) .. x_{ei}(R) \qquad x_e := x_{ei}(R), x_D .. x_D$$

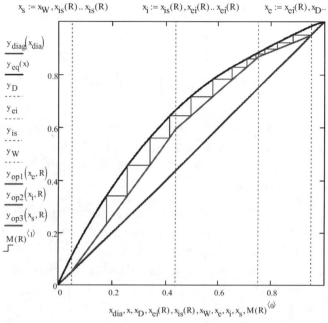

Using the assumed R, $x_{12} \neq x_W$. The correct R can be easily obtained by using *root* function (requiring that $x_{12} = x_W$): $\qquad R := root\big(M(R)_{N,0} - x_W, R\big) = 0.906$

The plot using this R is shown below:

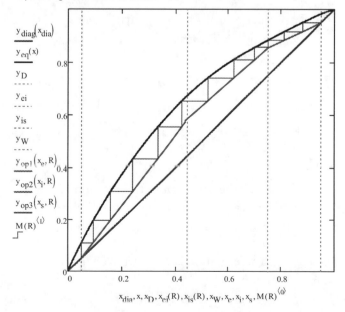

Problems

1. To represent the same function as the one discussed in Procedure 9.3, two other programming schemes are proposed:

$$f(x) := \begin{vmatrix} 0 & \text{if } x < 2 \\ 1 & \text{if } x \leq 5 \\ 3 & \text{otherwise} \end{vmatrix} \qquad f(x) := \begin{vmatrix} 3 & \text{if } x > 5 \\ 1 & \text{if } x \geq 2 \\ 0 & \text{otherwise} \end{vmatrix}$$

Do these programming schemes correctly represent $f(x)$ that we want? Explain.

2. Consider a convection heat transfer problem occurring in a smooth tube of constant wall temperature (T_s). The tube length and diameter are L and D, respectively. Water at a temperature of T_i enters the tube at a mass rate of m.

   a. Create a Mathcad program for calculating the exit water temperature (T_o) and the heat transfer rate (q). The water properties as functions of temperature are given as inputs.

   b. Apply the program to calculate the water exit temperature and the heat transfer rate if $T_s = 342.15$ K, $T_i = 305.15$ K, $m = 0.25$ kg/s, $L = 4$ m, and $D = 40$ mm. The following water properties as functions of temperature are obtained from the fitting of data in the range of 273.15 to 373.15 K:

   $$k(T) = -0.496665 + 5.945727 \times 10^{-3} T - 7.482665 \times 10^{-6} T^2$$
   $$\rho(T) = 246.631798 + 6.658227T - 0.018487T^2 + 1.541764 \times 10^{-5} T^3$$
   $$\mu(T) = 0.403337 - 4.638597 \times 10^{-3} T + 2.011653 \times 10^{-5} T^2$$
   $$\qquad - 3.890376 \times 10^{-8} T^3 + 2.827235 \times 10^{-11} T^4$$
   $$cp(T) = 1.8704258 \times 10^5 - 2.7097467 \times 10^3 T + 16.0500521 T^2$$
   $$\qquad - 0.0474905 T^3 + 7.0174807 \times 10^{-5} T^4 - 4.1406453 \times 10^{-8} T^5$$

   where T is the temperature [K], k is the thermal conductivity [W/(m·K)], ρ is the density [kg/m³], μ is the viscosity [N·s/m²], and cp is the specific heat [J/(kg·K)]

   c. Plot T_o as a function of m in the range of $0.01 \leq m \leq 0.5$.

   d. Apply the program to determine L if the desired T_o is 330 K. All the other parameters are the same as those in part (b).

   The flow chart shown on the next two pages can be used to create the program needed:

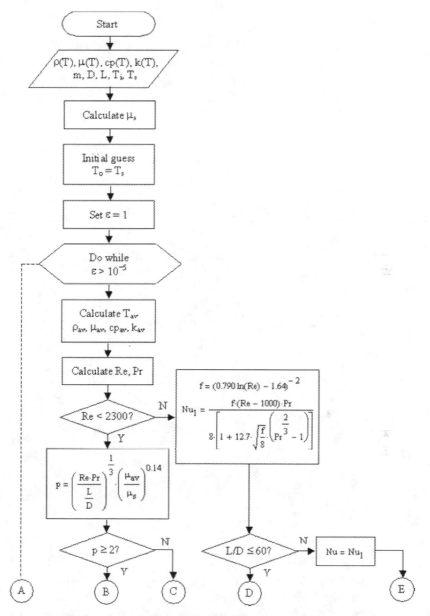

In the flow chart, subscript s refers to the property evaluated at T_s, subscript av refers to the average property evaluated at the average water temperature: $T_{av} = (T_i + T_o)/2$, Re is the *Reynolds* number, Pr is the *Prandtl* number, Nu is the *Nusselt* number, h is the convection coefficient, and ε is the approximate relative error. At the beginning of the program, T_o is guessed to calculate T_{av}. The new value of T_o is then calculated after h is obtained and used as a better guess to calculate a better T_{av}. The process is repeated until the new value of T_o is very close to the guessed T_o, i.e., until $\varepsilon \leq 10^{-5}$.

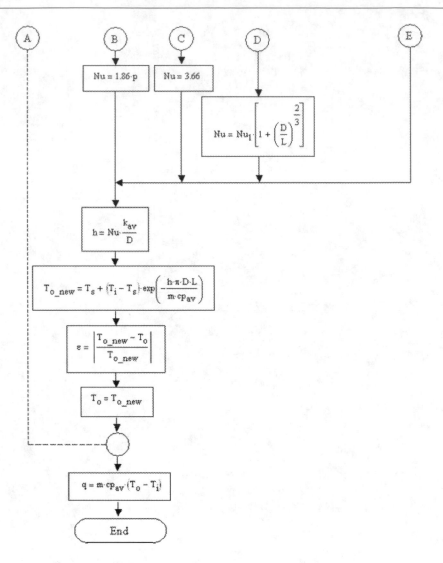

3. In the design of heat exchanger network, an important task is to determine the minimum heating load, the minimum cooling load, and the pinch temperature. To perform this, the first step is to determine the temperature intervals of the hot and cold streams associated with the chosen minimum driving force of ΔT. This minimum driving force is also referred to as the minimum approach temperature.

Suppose that the inlet and outlet temperatures of hot streams are stored in a vector T_1 and the inlet and outlet temperatures of cold streams are stored in a vector T_2. Of course, the number of rows of vector T_1 or T_2 is twice as many as the number of hot or cold streams. Also assume that no phase change is involved.

a. Develop a program to obtain a vector (T_h) containing tempera-
    tures that define the temperature intervals of the hot streams.
    This vector is created by merging of vectors T_1 and ($T_2+\Delta T$) and
    sorting the elements in ascending order. Any identical elements
    must be merged. Also, any element of vector ($T_2+\Delta T$) that is out-
    side the temperature range in vector T_1 must be discarded. Note:
    Use *sort* function to sort the elements in ascending order.

b. Develop another similar program to obtain a vector (T_c) contain-
    ing temperatures that define the temperature intervals of the cold
    streams. This vector is created by merging of vectors T_2 and
    ($T_1-\Delta T$) and sorting the elements in ascending order. Similarly,
    any identical elements must be merged and any element of vector
    ($T_1-\Delta T$) that is outside the temperature range in vector T_2 must be
    discarded.

c. Apply the two programs developed in parts (a) and (b) for the fol-
    lowing specific problem with $\Delta T = 10°F$:

Stream #		$F \cdot cp$ [Btu/(hr·°F)]		T [°F]
1	Hot stream	1000	inlet	250
			outlet	120
2	Hot stream	4000	inlet	200
			outlet	100
3	Cold stream	3000	inlet	90
			outlet	150
4	Cold stream	6000	inlet	130
			outlet	190

Thus, the vectors T_1 and T_2 are:

$$T_1 := \begin{pmatrix} 250 \\ 120 \\ 200 \\ 100 \end{pmatrix} \qquad T_2 := \begin{pmatrix} 90 \\ 150 \\ 130 \\ 190 \end{pmatrix}$$

4. After the temperature intervals of the hot and cold streams have been
    determined in Problem 3, the minimum heating and cooling loads
    and the pinch temperature can be calculated.

a. The cumulative heat available (cumulative enthalpy) in all of the
    hot streams is first calculated as a function of temperature defin-
    ing the temperature intervals of the hot streams, i.e., temperatures
    in vector T_h (see Problem 3). The enthalpy corresponding to the

coldest temperature of any hot stream ($T_{h,0}$) is set as the reference ($H_{h,0} = 0$). The heat available in temperature interval i is calculated as follows:

$$H_{h,i} = \sum_j F_{h,j} c_{ph,j} \left(T_{h,i} - T_{h,i-1} \right) \qquad i \neq 0$$

where $T_{h,i-1} \leq T \leq T_{h,i}$ is the temperature interval i, $F_{h,j}$ is the mass flow rate of hot stream j, $c_{ph,j}$ is the specific heat of hot stream j, and the summation is over all hot streams involved in the interval. The heat available is then cumulated as we move to higher temperature intervals. Thus, the cumulative enthalpy at $T_{h,i}$ is calculated from:

$$H_{h\_cum,i} = \sum_{j=0}^{i} H_{h,j}$$

Develop a program to obtain the cumulative heat available (cumulative enthalpy) at temperatures defining the temperature intervals of the hot streams.

b. Develop another program to obtain the cumulative heat requirement (cumulative enthalpy) at temperatures defining the temperature intervals of the cold streams, i.e., temperatures in vector T_c (see Problem 3). The enthalpy corresponding to the coldest temperature of any cold stream ($T_{c,0}$) is $H_{c,0}$ and considered as input, which needs to be determined later. The heat requirement in temperature interval i is calculated as follows:

$$H_{c,i} = \sum_j F_{c,j} c_{pc,j} \left(T_{c,i} - T_{c,i-1} \right) \qquad i \neq 0$$

where $T_{c,i-1} \leq T \leq T_{c,i}$ is the temperature interval i, $F_{c,j}$ is the mass flow rate of cold stream j, $c_{pc,j}$ is the specific heat of cold stream j, and the summation is over all cold streams involved in the interval. The cumulative enthalpy at $T_{c,i}$ is calculated from:

$$H_{c\_cum,i} = \sum_{j=0}^{i} H_{c,j}$$

c. By setting the value of $H_{c,0}$ to 1×10^5 Btu/hr, apply the two programs to the specific problem described in Problem 3(c). Plot the cumulative heat available and requirement on the same plot as a function of temperature defining the temperature intervals and determine the minimum temperature difference between the two curves. This minimum temperature difference is in fact the minimum approach temperature $\Delta T (= T_h - T_c)$ for the assumed $H_{c,0}$.

d. Determine the correct value of $H_{c,0}$ such that the minimum approach temperature $\Delta T = 10°F$, as stated in Problem 3(c). The minimum cooling load is the correct $H_{c,0}$, the minimum heating load is $H_{c\_cum} - H_{h\_cum}$, and the pinch temperature is defined as $(T_h + T_c)/2$. H_{c_cum} is the cumulative heat requirement at the highest temperature of any cold stream and H_{h_cum} is the cumulative heat available at the highest temperature of any hot stream.

5. A continuous distillation unit, i.e., a tray column with a partial reboiler and a total condenser, is to be designed to separate ethanol and water at atmospheric pressure. The feed, which is introduced into the column as liquid at its bubble point ($q_F = 1$), enters the column at a rate of 100 kgmol/hr (F) and contains 30 mol% alcohol (z_F). The distillate is to contain 80 mol% alcohol (x_D) and the bottoms product is to contain 4.5 mol% alcohol (x_W). The reflux ratio (R) is 2 and McCabe-Thiele method can be used.

The equilibrium data are as follows:

x	0	0.02	0.05	0.1	0.2	0.3	0.4	0.5
y	0	0.192	0.377	0.527	0.656	0.713	0.746	0.771

x	0.6	0.7	0.8	0.9	0.94	0.96	0.98	1.0
y	0.794	0.822	0.858	0.912	0.942	0.959	0.978	1.0

where x and y represent the mole fractions of alcohol in the liquid and vapor phases, respectively.

a. Create a function $y_{eq}(x)$ that represents the equilibrium data by using cubic spline.

b. Determine the molar flow rates of distillate (D) and bottoms (W) from the following mole balance:

$$F = D + W$$
$$Fz_F = Dx_D + Wx_W$$

c. Since the process only has one feed, the column has only two sections, i.e., enriching and stripping sections. The operating lines in these sections are given by

$$\text{Enriching:} \quad y = \frac{R}{R+1}x + \frac{x_D}{R+1}$$

$$\text{Stripping:} \quad y = \frac{L_s}{V_s}x - \frac{Wx_W}{V_s}$$

where the molar flow rates of liquid (L_s) and vapor (V_s) in the stripping section can be calculated from:

$$L_s = L_e + q_F F \qquad L_e = RD$$
$$V_s = V_e - (1 - q_F)F \qquad V_e = L_e + D$$

Create a Mathcad program, similar to that in Example 3 of Example set 9.1, to calculate mole fractions of alcohol in the liquid and vapor streams. The stage calculation steps off from (x_0, y_1) $[=(x_D, x_D)]$ until x_W is reached in the stripping section.

d. Plot the xy diagram containing the equilibrium curve, the diagonal line, the operating lines, and the stages for this problem.

Index